X·over 3 Pro

Passive crossover network design software for Microsoft® Windows®

User Manual

Fifth Edition (for version 3.0.18 and later)

by D.E.Harris

X•over 3 Pro User Manual, 5th Ed

ISBN-10: 1494773414
ISBN-13: 978-1494773410

Warning and Disclaimer
This book is designed to provide information about the X•over 3 Pro computer program (version 3.0.18 and later) for speaker designers. Every effort has been made to make this book as complete and as accurate as possible, but no warranty of fitness is implied.

The information is provided on an "as is" basis. The author and publisher shall have neither the liability nor responsibility to any person or entity with respect to any loss or damages arising from the information contained in this book.

Neither the X•over 3 Pro software nor a license to use the software are included with this book. The software and license to use the software must be purchased from its manufacturer, Harris Technologies, Inc., or one of the manufacturer's authorized distributors. See the manufacturer's website at www.ht–audio.com for more information.

Technical Support
Technical support for the X•over 3 Pro software is not available from the author of this book. All technical support requests should be sent to the software manufacturer, Harris Technologies, Inc. (Harris Tech). Harris Tech provides support two ways: General support topics are available on their website at www.ht–audio.com. These topics answer many common questions. Support is also available via email to support@ht–audio.com. Support is available only in the English language.

How to contact Harris Tech:
Email: support@ht–audio.com (technical questions)
sales@ht–audio.com (sales questions)
Website: www.ht–audio.com

Contents

X•over Pro Reference

Getting Started

This *User Manual* will show you how to use X•over Pro to design passive crossover networks, filters, driver impedance equalization networks and L-pads. **Please refer to the separate *Installation Instructions* included with the software if you have not yet installed the program.**

What is X•over Pro?

X•over Pro is an innovative passive network design program. It can help you design a 2-way or 3-way passive crossover network, a high-pass, band-pass or low-pass filter, an impedance equalization network, an L-pad and a series or parallel notch filter.

X•over Pro helps you design passive networks in two ways: 1) It calculates the component values for the desired network. 2) Using the driver and box Thiele-Small parameters, it models the system response of the network and speaker. This latter capability is very valuable when you want to see what happens when one or more component values are changed.

Calculating Network Component Values

The capabilities of X•over Pro automatically scale to match the number of driver parameters that are entered. For example, the components of a basic crossover network can be calculated with just the impedance of each driver. With the driver Thiele-Small parameters, the components of an impedance equalization network can also be calculated for each driver. Add the box parameters and the woofer's impedance equalization network can also compensate for the box resonance.

Modeling the System Response

Including the Thiele-Small parameters of the drivers in a design enables X•over Pro to model the "small-signal" response of the system. It is the response of the "system" because it depicts the whole speaker including the crossover network, impedance equalization networks, L-pads, drivers and box (if box parameters are included). It is called the "small-signal" response because it examines the network and speaker at small or low power levels. X•over Pro models the response with the following performance graphs:

- Normalized Amplitude Response (often referred to as the "frequency response").
- System Impedance Response.
- Phase Response.
- Group Delay.

Features

User Interface

- X•over Pro employs a modular user interface that remembers user settings. Extensive use of tab sheets help to concentrate more information into fewer windows.
- The program has a "real-time" feel. Changes made to a design are immediately reflected in other property windows and displayed in a mini "thumbnail" graph in the main window. A large schematic of the network is automatically updated when appropriate.
- Many default settings can be adjusted by the user. For example, the user can select from four different schematic layout options and whether the context-sensitive balloon help is turned on or off.
- English or metric units can be used (or a combination of both).
- An extensive, illustrated on-screen manual and help system are provided.

Crossover Network / Filter Design

- Design two-way and three-way crossover networks or separate low-pass, band-pass and high-pass filters. Choose from 1st, 2nd, 3rd and 4th-order topologies.
- Bessel, Butterworth, Chebychev, Gaussian, Legendre, Linear-Phase and Linkwitz-Riley filter types are available for two-way crossovers and separate low-pass and high-pass filters.
- All-Pass Crossover (APC) and Constant-Power Crossover (CPC) filter types are available for three-way crossovers and separate band-pass filters.
- Perform small-signal analysis when the driver Thiele-Small parameters are entered.
- The user can change any component value and see the effects in the graphs (the mini "thumbnail" graph updates in real-time). *Note: The graphs require driver T-S parameters.*
- Design impedance equalization networks for both the drivers and the box. Design L-pads to equalize the level of one or more drivers.

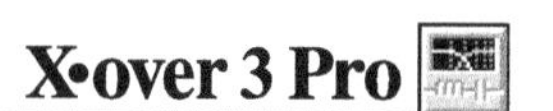

Graphs

Note: The graphs are operational when the driver Thiele-Small parameters are entered.

- Examine the performance of a design with four graphs: 1) normalized amplitude response; 2) system impedance; 3) phase response; and 4) group delay.
- The graphs show the total system response, including the response of the passive crossover network or filters, impedance equalization networks, L-pads, drivers and box. The measured acoustical response of the driver and listening environment can also be included. *Note: The latter feature requires BassBox Pro.*
- Two graph modes are available. For VGA resolutions, a single graph window displays graphs one at a time. For SVGA and XGA resolutions, a separate window with two size options is available for each graph so graphs can be viewed simultaneously.
- The graphs can display the estimated piston band on-axis amplitude rise of a driver and the estimated diffraction response shelf (of select box shapes) that can arise from the front-panel circumference of the box. *Note: The latter feature requires BassBox Pro.*
- Other options include two vertical scales, seven graph memories, a cursor and a Graph Properties window. The color and width of plot lines and the grid intensity are all adjustable. A graph can also be exported via the Windows clipboard for use in other programs such as word processors and page layout programs.

Parameters

- Both Thiele-Small and electromechanical parameters are used and can be entered manually or imported from a CLIO measurement system, a DATS (Dayton Audio Test System) or WT3 (Dayton Audio Woofer Tester 3), a LAUD version 3.12 (or later) measurement system, or the WT2 (Smith & Larson Audio Woofer Tester 2) measurement system. Parameters may also be imported from other BassBox 6 (Pro or Lite) "bb6" design files.
- Open back driver (woofer) and box parameters can be imported from a BassBox Pro or BassBox Lite speaker design file. If available, acoustic data about both the driver and external environment can also be imported and included in the performance graphs of X•over Pro. *Note: The driver acoustic data can be edited from within X•over Pro; however, the external acoustic environment cannot be edited from within X•over Pro.*
- An "Expert Mode" includes a self-analyzing feature for driver parameters. This feature automatically checks the parameters and places a green indicator beside the correct ones, a yellow indicator beside the marginal ones, a red indicator beside the incorrect ones and a grey indicator beside the ones that it is unable to test. The test sensitivity can be adjusted.
- Dual voice coil drivers are also accommodated with separate parameters for individual, parallel and series voice coil wiring.
- "Net" driver parameter values are displayed for multiple driver designs. These designs can select from three types of mechanical configurations (standard, isobaric and bessel) and

four types of electrical configurations (parallel, series, series-parallel and separate). Standard and isobaric configurations can also be push-pull.

- The external series resistance and amplifier source resistance can be entered and their effects included in the graphs.
- An extensive driver database with woofers, midrange drivers and tweeters is provided. It includes the parameters of thousands of drivers and can be searched by manufacturer, model name, driver parameters or certain box parameters. The user can add, edit or remove drivers from the database. The driver database is compatible with BassBox Pro.
- Basic box parameters can be manually entered into X•over Pro so that a comprehensive impedance equalization network can be designed to equalize the combined driver and box response.

Acoustical

- Acoustic measurements of the drivers can be imported from many popular measurement systems (including Brüel & Kjaer, CLIO, IMP, LMS, MLSSA, OmniMic, Sample Champion, Smaart, TEF®-20 and TrueRTA).

Printouts

- Printouts can include a list of components and their values, a schematic, lists of driver and box parameters and four graphs.
- A custom logo or graphic can be imported and printed in the title block. Many popular bitmap file formats are supported.

Miscellaneous

- A Filter Component Resistance Estimator is available to estimate the ESR (equivalent series resistance) of capacitors and DCR (DC resistance) of inductors.
- A Parallel-Series Value Calculator is available to calculate the net value of up to four different parallel-wired or series-wired components.
- A Color Value Decoder is available to decode the value from a component's color stripes.
- A Notch Filter Designer is available to calculate the component values for a simple series or parallel notch filter.
- Older X•over 2.1, 2.0 and 1.0 files can be opened.

System Requirements

X•over 3 Pro requires Microsoft® Windows® and has broad compatibility with 32-bit versions of Windows 8*, 7, Vista, XP, 2000, NT4, Me and 98. It also runs under many 64-bit versions of Windows. The program requires a minimum of 31 MBytes of free hard disk space and a CD-R compatible CD-ROM, DVD or Blu-ray drive. Also, the Arial and Symbol TrueType fonts must be installed (both fonts are a standard feature of Windows). For best results, a single Windows user account with administrator privileges should be used to install and run the program.

*Windows RT, the tablet version of Windows 8, is not supported.

User Requirements

X•over 3 Pro is a versatile computer program that can serve a wide range of speaker designers. Yet, it is no substitute for experience and creative insight when it comes to passive crossover network and filter design. Some basic familiarity with passive filters and their components is advised.

It is this author's hope that this *User Manual* will similarly serve a wide range of X•over 3 Pro users. It includes general information, hints and tips for new users and in-depth information about the operation of X•over Pro for experienced designers. However, this manual does not pretend to be a tutorial on crossover network or filter design nor a tutorial in the use of Microsoft Windows. For additional information on crossover design, please refer to the Suggested Reading list in Appendix E. For additional information on Windows, please refer to Microsoft and/or your computer vendor.

Starting X•over Pro

X•over Pro can be started from the Windows Start menu. If the program was installed using the default settings, the menu path is "Start > Programs > HT Audio > X•over 3 Pro" as shown below:

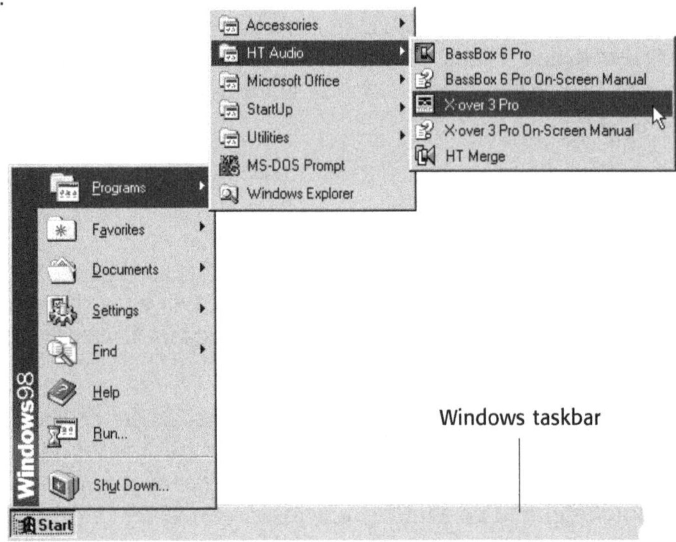

The first time that X•over Pro is run after being installed, a "Welcome" window will appear (shown on the top of the next page). It will walk you through the following steps:

- Select a default schematic style.
- Select English or metric units as the default.
- Select "Normal" or "Expert" mode for the driver parameters.
- Read selected portions of the on-screen manual to learn about the program.

Use the "Next" button at the bottom of the window when prompted to advance through each of these steps. A "Finish" button will appear at the end. Clicking (🖱) it will save the configuration and the "Welcome" window will close.

The next time you run X•over Pro, its standard title window will appear (shown on the bottom of the next page).

Notes: If you previously used the "Cancel" button to close the "Welcome" window without saving an initial configuration, then the "Welcome" window will continue to appear in place of the program's standard title window. On the other hand, if the "Finish" button was clicked but the "Welcome" window continues to appear anyway, it may be necessarly to give X•over Pro administrator privleges every time you launch it. This can be done by right-clicking (🖱) on the program and selecting "Run as administrator".

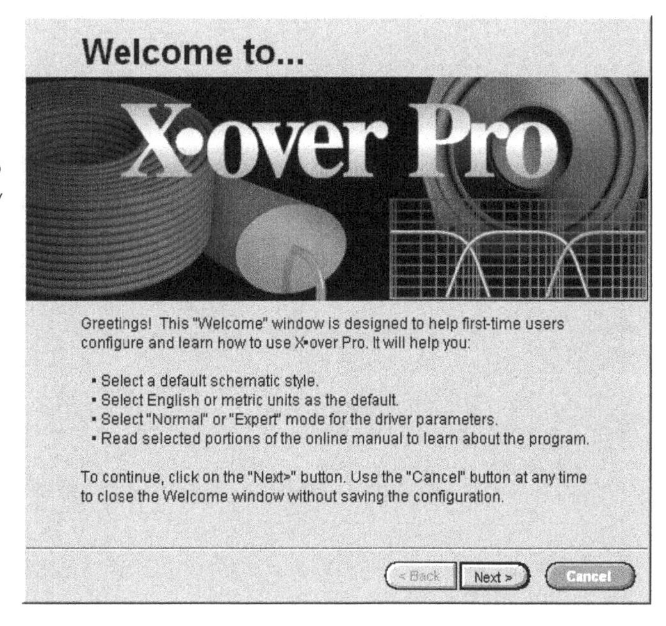

X•over Pro "Welcome" window

The title window (shown below) includes the program version number, copyright notice, licensee name, company name and program serial number. To open this window later while the program is running, select "About X•over Pro" from the Help menu. Notice the two buttons located at the bottom:

- **Open Design Window** – opens the X•over Pro main window shown on the next page. From it you can begin a new design or open an existing design.
- **Close X•over** – stops the loading of X•over Pro and closes the title window.

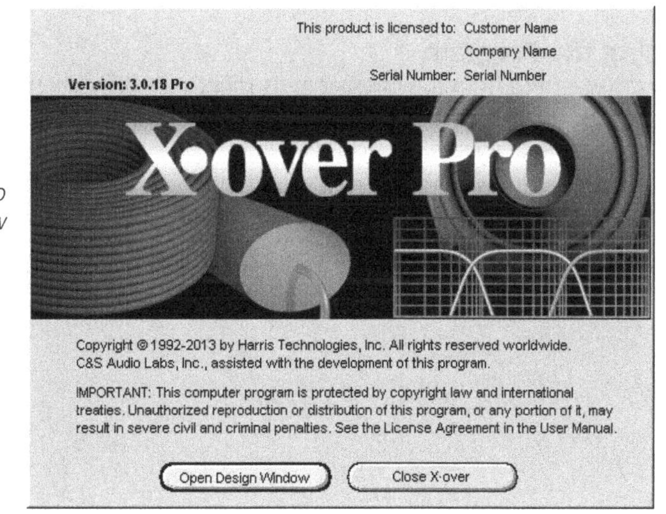

X•over Pro title window

The main X•over window is shown below. It is empty when X•over Pro is first opened:

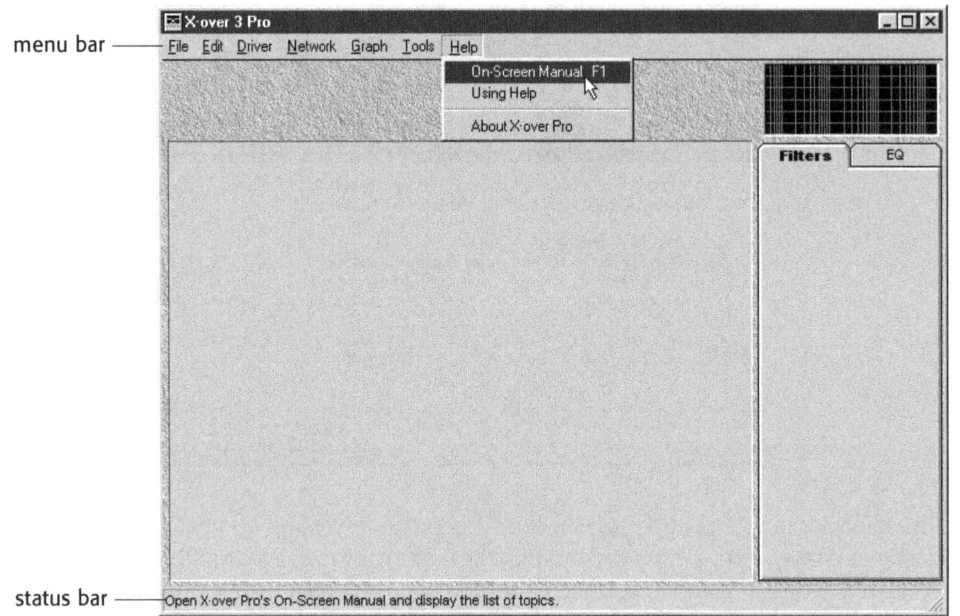

X•over Pro main window

Like most Windows applications, the main window has a menu bar where many commands can be found. For example, use the File menu to begin a new design or open an existing design. Use the Help menu to open the on-screen manual.

Quitting the Program

To close X•over Pro from the title bar of its main window, click the Close button at the far right end of the title bar. To close X•over Pro from its menu bar, select "Quit" from the File menu or use the keyboard shortcut Ctrl+Q. The keyboard shortcut works only when the program has the focus. If you have made changes to a design that have not been saved, you will be given an opportunity to save them before the program terminates.

Overview

The heart of X•over Pro is its main window (shown below) which includes several filter controls, a schematic and a list of network components and their values.

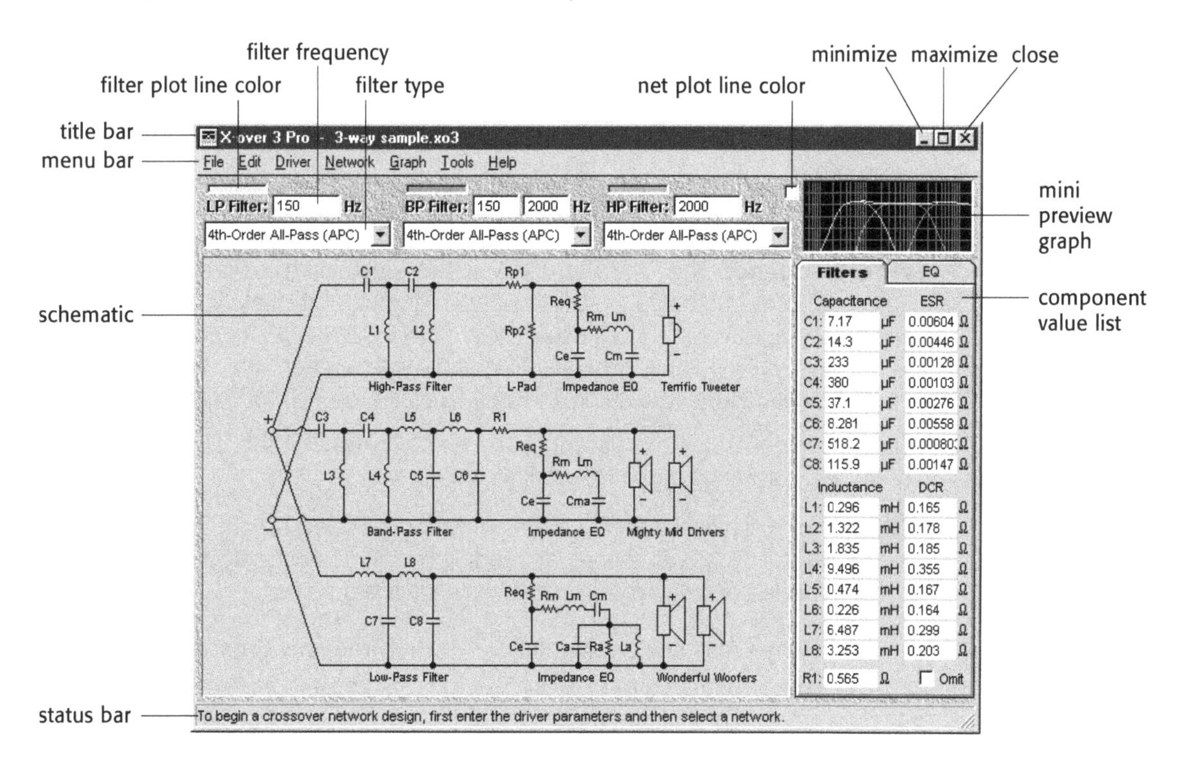

Like most Windows applications, it has a menu bar where most of the commands can be found. It can also be resized and the size and scale of the schematic will be automatically adjusted to fit. However, there is a minimum size limit–the main window cannot be resized to a size that is smaller than VGA resolution (640 x 480 pixels).

The major components of the main window are:

Title bar Lists the program name and the file name of the current design file (if one is open). It also includes the Minimize, Maximize and Close buttons. Like other Windows applications, the color of the title bar also indicates which window has the focus.

Menu bar The menus of the menu bar contain many of the commands for the program. Many commands also have keyboard shortcuts which are listed beside them in the menus. A list of keyboard shortcuts is also provided in Appendix A.

Filter plot line color There are three Filter Plot Line Color controls which serve as both indicators and buttons. They display the plot line color of the individual low-pass, band-pass and high-pass filters that make up a crossover network. By clicking (🖱) on them the color will advance one at a time through a 12-color palette with each single click. The first ten colors of the palette are fixed (red, orange, yellow, greenish-yellow, green, cyan, blue, magenta, white and light grey). The last two colors are "custom" colors and you can make them any color you'd like. You can set the default plot line colors and you can select the custom colors with the "Graph" tab of the Preferences window.

Changing a Filter Plot Line Color setting will not change the color of plot lines that already exist in the graph windows. It will only affect future plotting.

Filter frequency The Filter Frequency is the same as the crossover frequency for two adjacent filters if both use the same Filter Frequency and Filter Type. If they are the same, the crossover point will usually vary from –3 to –6 dB, depending on the Filter Type selected. The actual crossover frequency will vary from the Filter Frequency value if dissimilar Filter Frequencies and/or Filter Types are chosen for adjacent filters.

The Filter Frequencies of adjacent filters can be linked with the "Link Adjacent Filter Frequencies" command in the Network menu.

Filter type Each filter has a drop-down list for the selection of the Filter Type. The available Filter Types will change depending on the network selection. The choices for two-way and three-way crossover networks are listed below:

Two-way Crossover Networks

1st-order	Butterworth (APC & CPC) Solen Split –6 dB
2nd-order	Bessel Butterworth (CPC) Chebychev Linkwitz-Riley (APC)
3rd-order	Butterworth (APC & CPC)
4th-order	Bessel Butterworth (CPC) Gaussian Legendre Linear Phase Linkwitz-Riley (APC)

Three-way Crossover Networks

1st-order	All-Pass (APC) Constant-Power (CPC) Solen Split –6 dB
2nd-order	All-Pass (APC) Constant-Power (CPC)
3rd-order	All-Pass (APC) Constant-Power (CPC)
4th-order	All-Pass (APC) Constant-Power (CPC)

When separate filters are selected instead of a two-way or three-way crossover network, a combination of the above Filter Types will be available. The low-pass and high-pass filters will use the two-way choices and the band-pass filter will use the three-way choices. See the "Filter Summary" in Chapter 1 of the *Crossover Network Designer's Guide* later in this manual for a description of each filter.

Net plot line color (Two-way and three-way crossover networks only.) The mini preview graph and two of the full size graphs provide a "net" response option to plot the combined response of each filter in the crossover network. This control serves both to indicate the color and provide a means of changing it. Simply click (🖱) on it to advance through the 12-color graph palette.

The net response of the mini preview graph and Normalized Amplitude Response graph can be calculated one of two ways, depending on the setting of the "ø" (phase) checkbox on the "Graph" tab of the Preferences window. When "ø" is turned on, the net response is calculated by summing both the magnitude and phase of the volume velocity of each filter. **Important:** It assumes that all drivers are carefully mounted with their acoustical centers located close together and aligned so that the direct sound waves of each driver arrive to the listener's ear at the same time at the crossover frequency. The driver mounting you choose can drastically alter the actual net acoustical response that you receive. This is one reason why the second net response option is provided. To use it, turn off the "ø" control and the net response will be calculated by summing only the magnitude of each filter. Phase cancellations will be ignored with this method.

Minimize button Removes X•over Pro from the desktop without closing it. It can be restored by clicking on the X•over Pro button on the Windows taskbar.

Maximize button Causes the main window to fill the screen. Clicking the same button again will restore the main window to its former size.

Close button Clicking the Close button has the same effect as selecting "Quit" from the File menu (Ctrl+Q). It will close the program. Before the program is closed, you will be given the opportunity to save unsaved changes to a design.

Mini preview graph Displays the normalized amplitude response of the system from 5 Hz to 20 kHz with a vertical scale of 9 dB/division. It replots automatically whenever the design is changed. In this way it provides real-time feedback of the network while you work on it. You can force it to replot by left clicking (🖱) on it. Right clicking (🖱) on it will display a popup menu containing some of the same options of the full-size graphs. Changing any of these options will affect all X•over Pro graphs.

Schematic The most prominent feature of the main window is the large schematic or circuit diagram depicting the crossover network, driver impedance equalization networks, L-pads and driver wiring configurations. It is automatically updated whenever the design is changed. Four styles are available (see the next page).

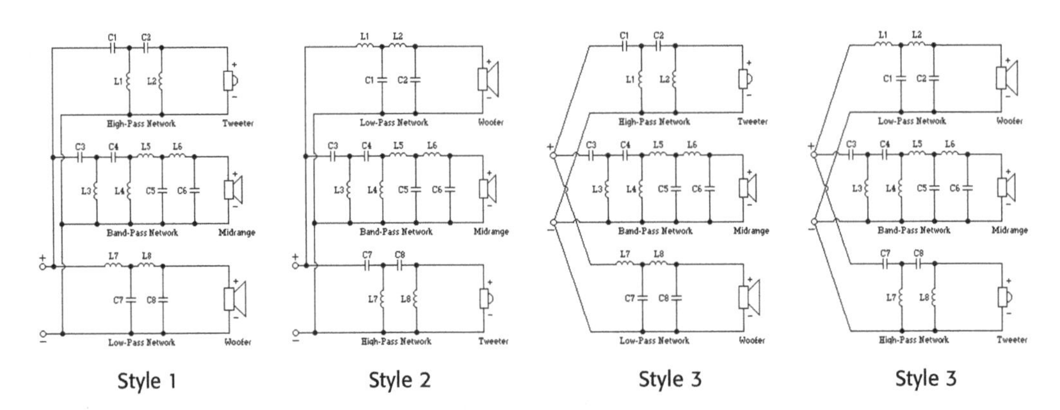

Style 1 Style 2 Style 3 Style 3

The styles differ in the way the input is drawn and whether the tweeter or the woofer is placed on top. Double-click (🖱🖱) anywhere on the schematic to toggle through each of the styles. The schematic will print the same way it is drawn on screen. The default style is set in the "General" tab of the Preferences window.

When multiple drivers are specified, there may be an occasion when the schematic is too wide to fit within its allotted area on the screen. When this happens, you can scroll the schematic left and right to view all of it. This is shown in the illustration below.

The mouse pointer changes to a left-right arrow to scroll the schematic.

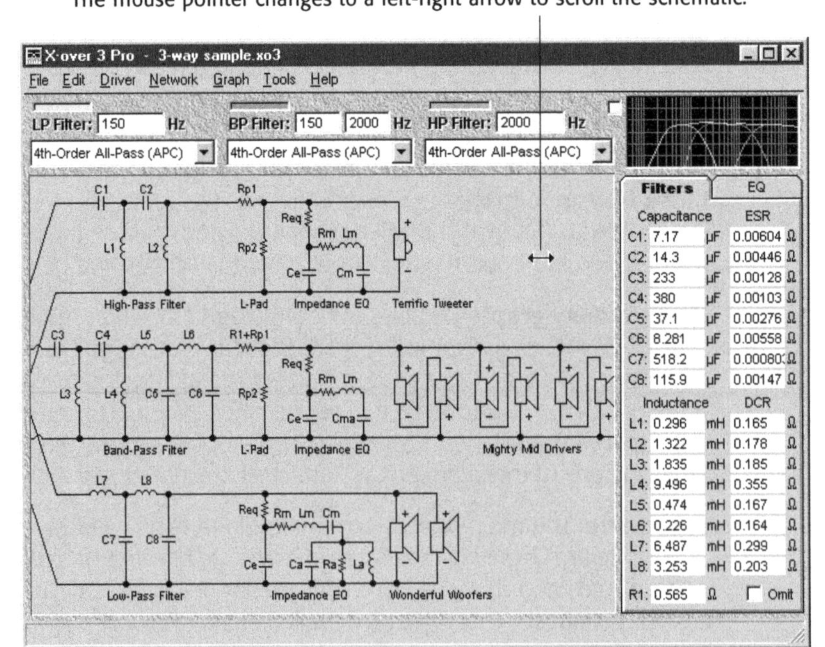

Notice that the mouse pointer changes to a left-right arrow when the left mouse button (🖱) is pressed to show that the schematic is ready to be scrolled. To move the schematic left and right, drag the mouse left and right over the schematic while the left button is held down.

Component value list A tab form is used to displays a list of all capacitors, inductors and resistors in the crossover network, impedance equalization networks and L-pads. This information is divided between two tabs (shown below). The "Filters" tab contains the components for the filters which make up the crossover network. The "EQ" tab contains the components for the driver impedance equalization networks and L-pads.

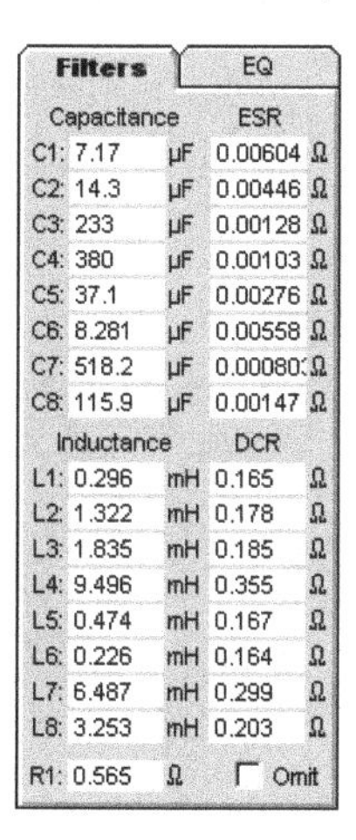

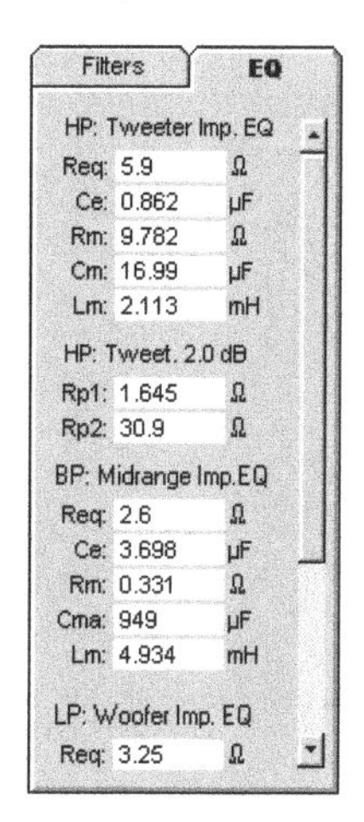

Notice at the bottom of the "Filters" tab that resistor R1 can be omitted if desired by checking the "Omit" checkbox. (R1 is used to attenuate the passband gain that is common with band-pass filters.) The band-pass filter component values are adjusted to reflect the presence or absence of R1.

The equivalent series resistance (ESR) and DC resistance (DCR) are required for the capacitors and inductors in the filters of the crossover network but not for the impedance equalization networks. X•over Pro will automatically estimate the ESR and DCR whenever possible. They can also be manually entered.

To estimate the ESR and DCR, the program must know what type of capacitor or inductor is used. This can be set with the Filter Component Resistance Estimator window. To open it, right-click (🖱) on any ESR or DCR input field in the component list or select the "Component Resistance Estimator" command from the Tools menu (keyboard shortcut Ctrl+E). Default settings are located on the "Parts" tab of the Preferences window.

Status bar Located at the bottom of the main window, the status bar displays tips, instructions or the status of a function.

Beginning a New Design

When the X•over Pro main window first opens (shown below) it is ready to begin a new design. If a design is already open, you can close it and begin a new one with the "New Design" command in the File menu (keyboard shortcut Ctrl+N).

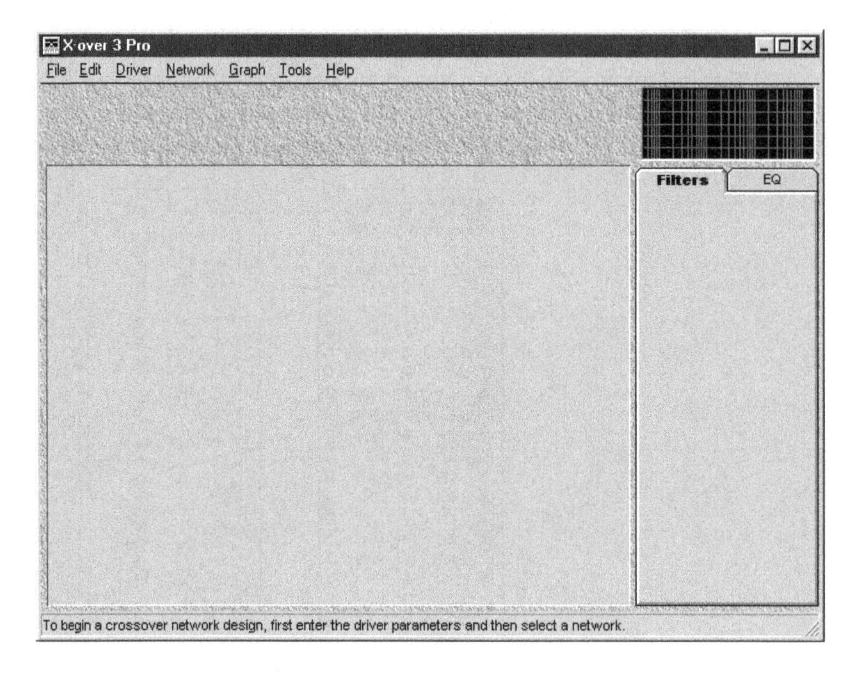

The design process usually follows these steps:

1. Enter the driver parameters (see the "Parameters" topic of Chapter 2 of the *X•over Pro Reference* later in this manual).
2. Enter the external resistance, if appropriate (see the "External R" topic in Chapter 2 of the *X•over Pro Reference*).
3. If needed, design an impedance equalization network for those drivers that require one (see the "Impedance EQ" topic of Chapter 2 of the *X•over Pro Reference*).
4. Select a crossover network or filter (see Chapter 3 of the *X•over Pro Reference*).
5. Evaluate the design with the performance graphs (see Chapter 5 of the *X•over Pro Reference*).
6. If needed, adjust the driver sensitivities with L-pads (see Chapter 4 of the *X•over Pro Reference*).
7. Prototype, test and adjust the network (see Chapter 7 of the *Crossover Network Designer's Guide* later in this manual).

Getting Help

X•over Pro provides four different help systems. Each one is described next:

Balloon Help This help system gets its name from the small message boxes which pop up like small text balloons whenever the mouse pointer (🖱) is paused over one of the many objects in the various windows of X•over Pro. A sample is shown below:

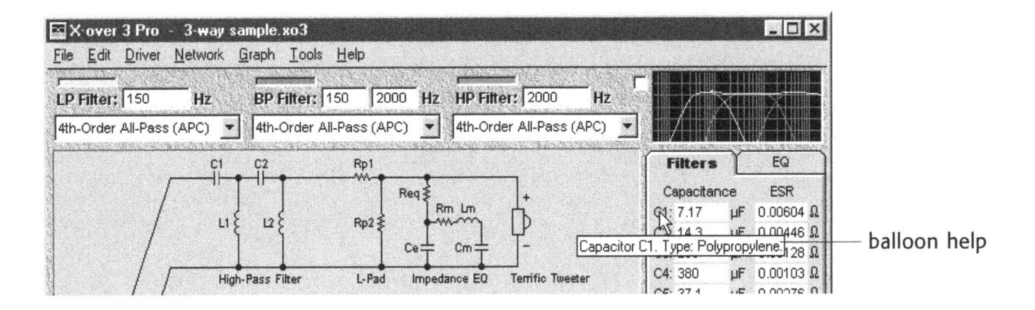

By the way, if you don't need to use balloon help, you can disable it. For details see Chapter 11 of the *X•over Pro Reference* section of this manual.

Status Bar A status bar is provided at the bottom of the main window. A description of each menu command is displayed in the status bar when a menu command is selected. It is also used to display tips, instructions and the status of some functions.

Context Sensitive Help Pressing the F1 or "Help" key will open X•over Pro's on-screen manual to a relevant topic. For example, pressing F1 while viewing the Filter Component Resistance Estimator window will open the on-screen manual to the topic on this subject (shown at right). Scroll through the topic to find information about a specific feature.

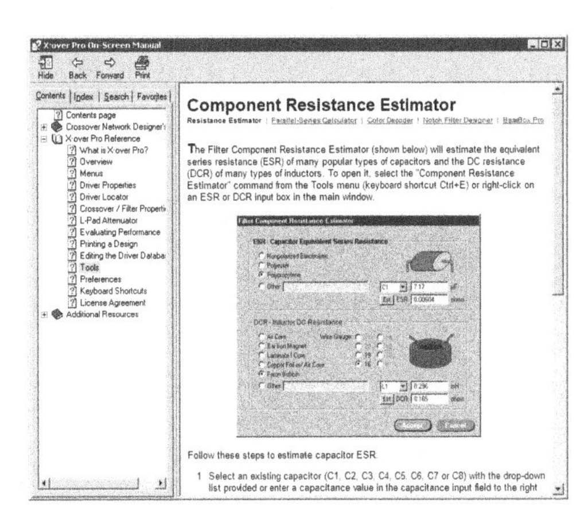

X•over Pro on-screen manual

On-Screen Manual Select "On-Screen Manual" from the Help menu of the main window to open it to its "Contents" page. From there you can select any of the major topics, including topics in the *Crossover Network Designer's Guide* section which are not accessible with context sensitive help.

The on-screen manual can be opened even when X•over Pro is not running. A direct link to it is included with X•over Pro in the Start menu as shown on page 12.

Technical Support

When the preceding help systems do not provide the assistance you need, then software support is available in two ways. First, *Technical Notes* are provided at the Harris Tech website and can be accessed 24 hours per day, 7 days per week. The *Technical Notes* answer some common questions. The web address to the X•over 3 Pro tech note index is:

www.ht–audio.com/pages/support/XoPSupport.html

Second, support is also available from Harris Tech via email at **support@ht–audio.com**. Support is available only in the English language. Please remember to include your X•over Pro serial number when you contact them.

Crossover Network Designer's Guide

This is the place to start if you are new to passive crossover network or filter design. This *Crossover Network Designer's Guide* begins by explaining the purpose and function of a crossover network. It answers questions like "Why does a speaker need a crossover network?" and "What is a 'passive' crossover network?" It describes the components of a crossover network and presents several associated topics.

Chapters

1 Crossover Basics

Most speakers contain three types of components: a box, a crossover network and two or more drivers as shown below:

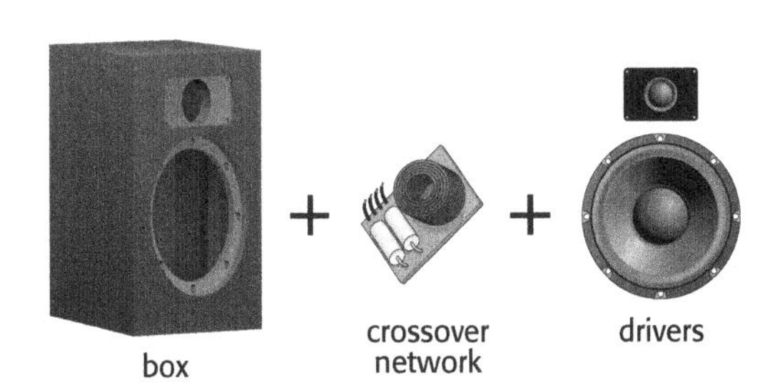

The box controls the speaker's low-frequency response and houses the drivers and passive crossover network. The crossover network divides the electrical audio signal between the drivers. The drivers convert the electrical audio signals into sound waves that we can hear. X•over Pro is concerned primarily with the crossover network but the crossover network cannot be designed in a vacuum–it is influenced by the drivers and the box.

The phrase "crossover network" is the popular name of a type of electrical circuit known as a "dividing network". The name is sometimes shortened to "xover" and this is where X•over Pro gets its name. X•over Pro can design a variety of crossover networks.

This chapter is divided into the seven topics listed below. They provide a basic overview of crossover networks, their purpose, components and design.

"Introduction" – Why a crossover network is needed and where it is located.
"Filters: the Building Blocks" – The filters that comprise a crossover network.
"Crossover Topology" – The circuit layout of a crossover network.
"Filter Topology" – The circuit layout of a filter.
"Filter Order" – A popular filter classification that describes its cutoff rate.
"Filter & Crossover Types" – Common types of filters and crossover networks.
"Filter Summary" – A brief description of each filter in X•over Pro.

1

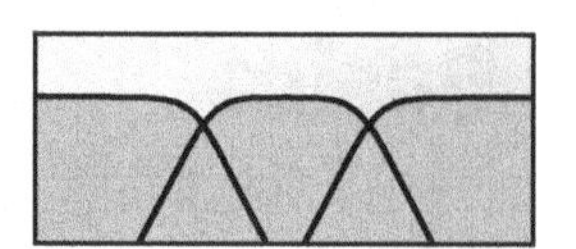

Introduction

Crossover networks "divide" the audio signal. Why? Why does the audio signal need to be divided and what does it mean to "cross over"? The answers to these questions begin with the drivers because they are the ultimate reason why a crossover network is needed.

Drivers

There are many different kinds of drivers but they all do basically the same thing: create sound waves. By far, the most common type of driver is the moving coil electrodynamic piston driver. It has a moving part called a diaphragm that acts like a piston to pump air and thereby create sound waves. A common diaphragm for a woofer is a paper cone. A common diaphragm for a tweeter is a fabric dome. The moving coil piston driver is the type that X•over Pro models so our discussion will focus on them.

Why do drivers come in so many different sizes as shown above? Because it is nearly impossible to make one piston driver that can produce uniform sound waves over the entire 20 Hz to 20 kHz frequency range of human hearing. To produce low frequencies a driver needs to have a large diaphragm and enough mass to resonate at a low frequency. To produce high frequencies a driver needs to have a small diaphragm with a low mass. Obviously, these requirements are in opposition so drivers are usually optimized to produce only a portion of the sound. This gives rise to multi-way speaker systems. In the illustration at right, a two-way speaker uses a tweeter for the high frequencies and a woofer for the low frequencies.

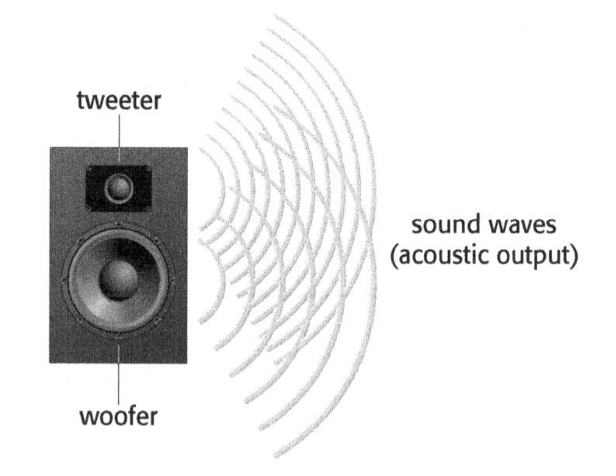

Tweeters are the smaller drivers since they produce the highest frequencies with the shortest wavelengths. Woofers are the largest drivers since they produce the lowest frequencies with the longest wavelengths. Are there other driver sizes? Yes, there are also midrange drivers of various sizes that reproduce middle frequencies between the tweeter and woofer. Midrange drivers are used in multi-way speakers with three or more driver sizes.

What happens if no crossover network is used in a two-way speaker? The tweeter and woofer would each be driven with a full 20 Hz to 20 kHz audio signal resulting in the following:

- The tweeter would not be able to handle much power and would be easily damaged. This is because the diaphragm of a driver must move farther as the frequency becomes lower and consequently the wavelengths become longer. Drive a typical piston tweeter with a low-frequency signal, like a bass guitar, and it will quickly fail.
- There would be many phase problems because the sound waves produced by the tweeter and woofer would overlap over the entire 20 Hz to 20 kHz range of the audio signal resulting in many "out-of-phase" cancellations and "in-phase" additions.
- The speaker would have decreased uniformity of direct sound coverage because the woofer would "beam" at high frequencies. This means that it would become more directional and sound loud on-axis and quiet off-axis.
- The tweeter would have increased distortion because of excessive diaphragm excursion at low frequencies. The woofer would have increased distortion because of "break-up" modes at high frequency resonances.

Crossover Network

Ideally each driver should operate only within its "optimum" frequency range. A tweeter should produce only high-frequency sound waves. A midrange driver should produce sound waves in the middle of the audio spectrum. A woofer should produce only low-frequency sound waves. Thus, a crossover network is required to divide the electrical audio signal into different frequency bands. When the audio signal is divided properly, the tweeter stops producing sound just as the midrange driver begins to produce sound. The midrange driver stops producing sound just as the woofer begins to produce sound. In this way, the job of creating sound waves is said to "cross over" from one driver to the next as the frequency content of the audio signal changes.

The picture on the next page illustrates this principle for a two-way speaker. A two-way speaker requires the electrical audio signal to be divided into two parts—a high-frequency part and a low-frequency part—before the signals reach the drivers.

1

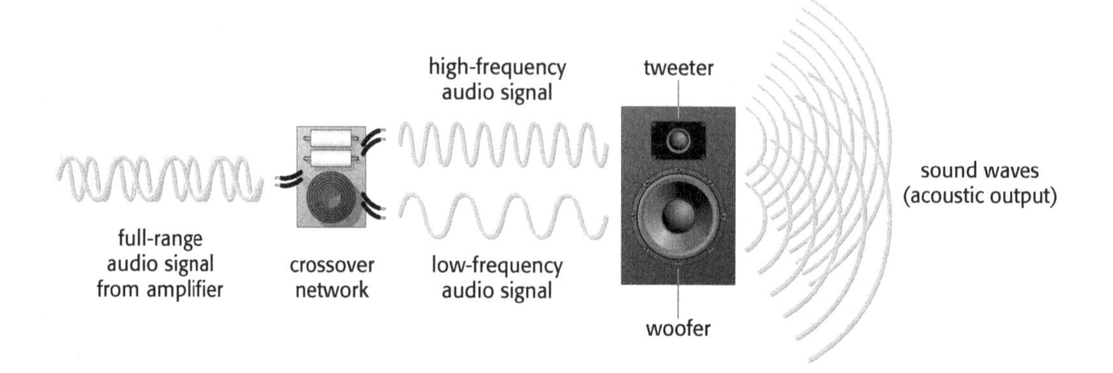

At some point, the high-frequency and low-frequency signals meet and overlap as shown below. The frequency where this occurs is called the "crossover frequency".

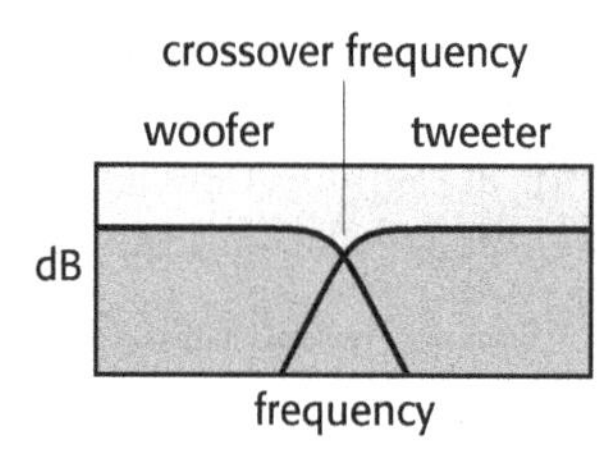

Normally the crossover frequency is adjusted so that the low-frequency and high-frequency parts of the signal have a –3 or –6 dB level at the crossover point. The exact level depends upon the type of filters chosen (see the "Filter Summary" later in this chapter). The goal is for the sound waves to sum in such a way that the resulting composite or "net" amplitude response is flat in the crossover region as if one driver were producing all of the sound. This is illustrated below:

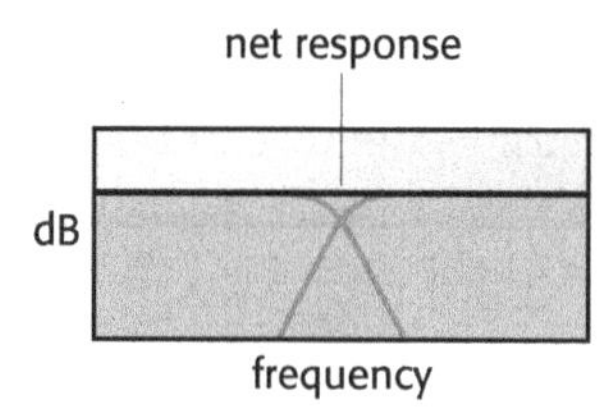

If the speaker has more than two size drivers the crossover network would also divide the audio signal into one or more additional midrange frequency bands.

Location

There are two general locations for a crossover network in the signal flow of the audio system: after the amplifier or before the amplifier. This is depicted below:

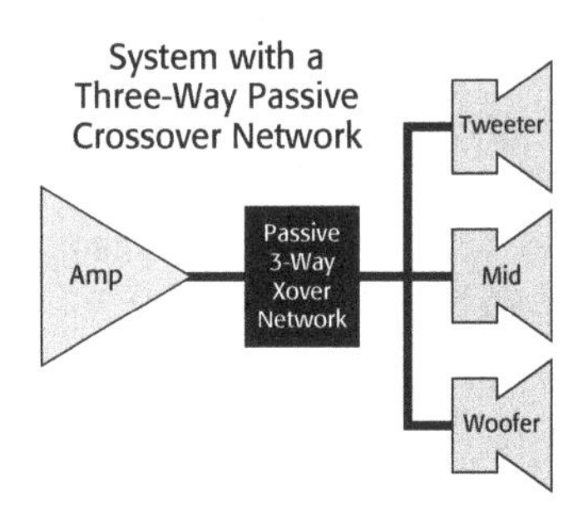

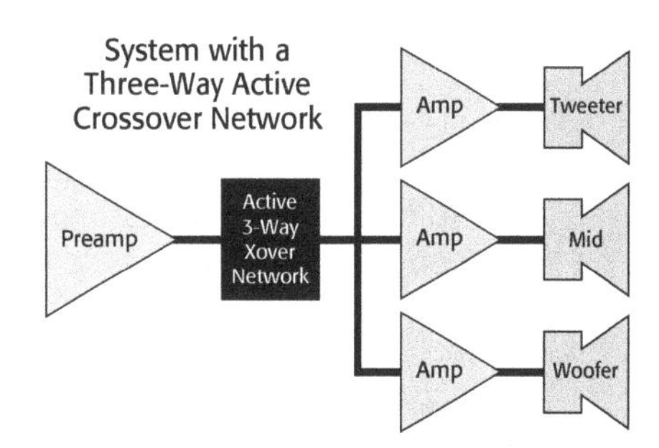

X•over Pro supports only one of these locations. Before discussing what X•over Pro expects, let's review some relevant points about each location:

"Passive" crossover network after the amplifier:

- Only one amplifier channel is required per speaker because the audio signal is divided after it has been amplified. This results in a lower overall cost for the audio system.
- The most commonly used location.
- Uses components that do not require an external power supply and so they are referred to as "passive" crossover networks.
- Uses large components that can handle the full power delivered to the speaker.
- Is very sensitive to the impedance response of the drivers.
- Can be mounted inside or outside of the speaker box. When a crossover network is mounted in or on the speaker box, it is considered a part of the speaker.
- Are often easier for the hobbyist to construct at home because a printed circuit board and power supply are not required.

"Active" crossover network before the amplifier:

- Requires a separate amplifier channel for each driver or crossover network filter because the audio signal is divided before it is amplified. This raises the overall cost of the audio system.
- Can produce higher fidelity and offer more adjustability.
- Uses components that require an external power supply and so they are referred to as "active" crossover networks. *Note: Passive crossover networks can also be used before the amplifier but this is rarely done because they are more difficult to design.*
- Uses smaller components since they are located "upstream" of the amplifier outputs.
- Is not affected by the impedance response of the drivers.

- Must be located between the preamplifier and power amplifier(s), usually in an equipment rack or cabinet. Because it is not located with the speaker, it is usually considered a separate component and not a part of the speaker.
- Are usually more difficult to construct because a printed circuit board and case (chassis) are often desired and an external power source is required.

X•over Pro can only design passive crossover networks and it assumes that they are located after the amplifier. The Thiele-Small parameters of the drivers (including the tweeters) are used to model the complex impedance of the "load" seen by each filter in the crossover network. This makes it possible for X•over Pro to predict the amplitude response, system impedance, phase response and group delay in the graphs.

Summary

A speaker is usually comprised of three parts: a box, a crossover network and one or more drivers. This "Introduction" topic discussed the following:

- A single size moving coil piston driver cannot accurately reproduce the entire 20 Hz to 20 kHz frequency range of human hearing and so most high-fidelity speakers use more than one size driver. This is why a crossover network is often needed.
- A crossover network divides the audio signal into separate frequency bands so that each driver receives the portion that it can best reproduce and so that overlap between the drivers is minimized.
- Crossover networks come in two varieties: the more common passive crossover network which is usually located after the amplifier and the more expensive active crossover network which is located before the amplifier(s).

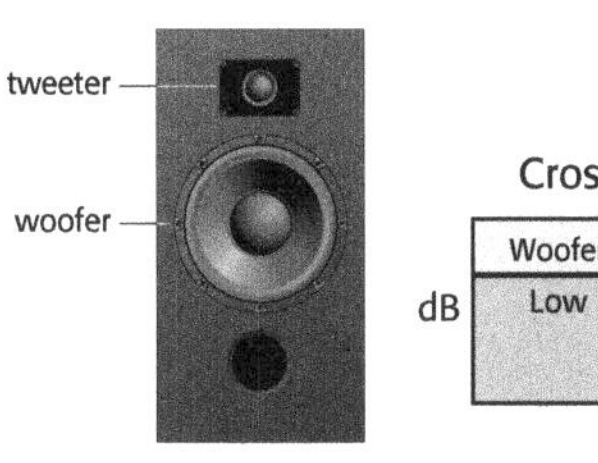

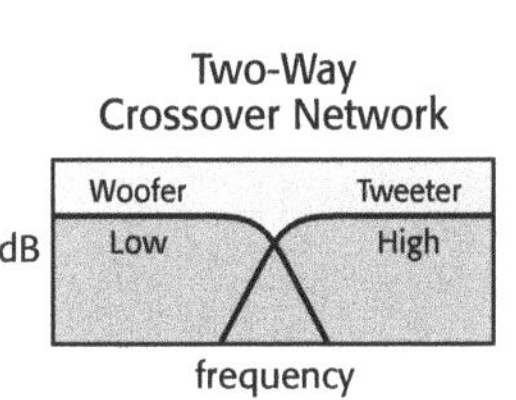

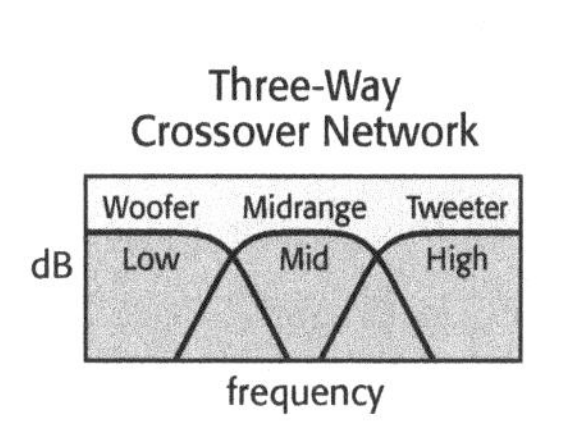

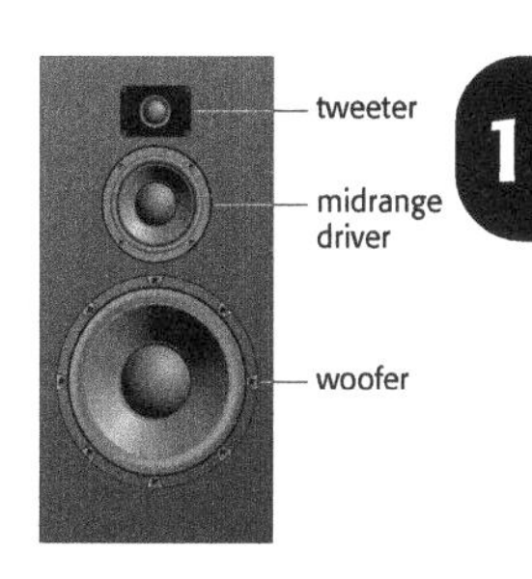

Filters: the Building Blocks

Filters are the "building blocks" of a crossover network—every crossover network has at least two of them. A simple two-way crossover network has a low-pass filter and a high-pass filter. Their purpose is to divide the electrical audio signal so that the woofer receives only the low frequency portion of the signal and the tweeter receives only the high frequency portion of the signal. The name "low-pass" means that the filter allows only low frequencies to pass through it. High frequency signals are filtered out. Conversely, a "high-pass" filter allows only high frequencies to pass through it because it filters out low frequency signals.

A three-way crossover network also includes a band-pass filter. A "band-pass" filter combines elements of both a low-pass and high-pass filter to allow a middle portion or a band of the audio signal to pass. It does this by filtering out both unwanted low frequency signals and unwanted high frequency signals. The response of all three filters is shown below:

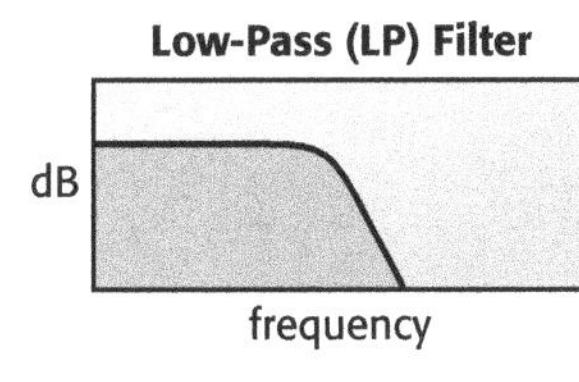

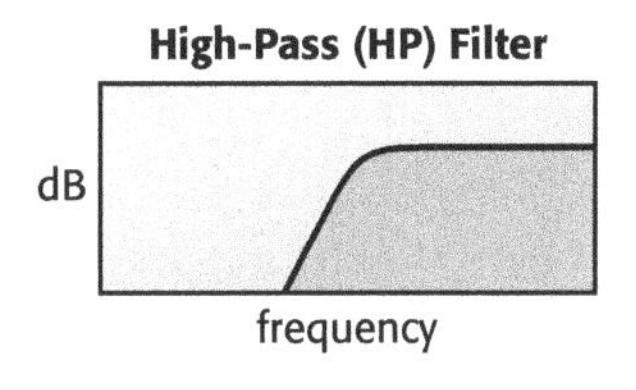

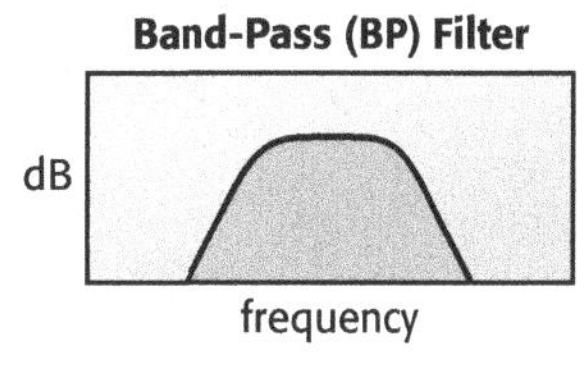

The portion of the audio signal that is passed by a filter is called its "passband" and the portion of the signal that is attenuated by a filter is called its "stopband." When the various filter elements of a crossover are added together, they should produce an overall smooth or flat amplitude response as shown at right:

Ideally, a crossover network should sum to a flat response.

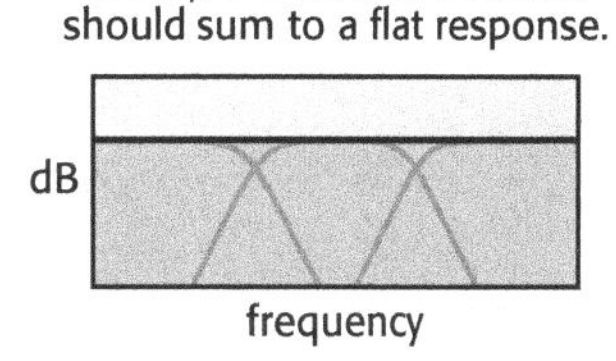

However, the ultimate goal is not a smooth electrical response at the output of the crossover network but rather a smooth acoustic response from the speaker. Unfortunately there are many factors which can prevent this. Some are internal to the crossover network and its design and some are external. A list of these factors is provided on the next page.

1

Factors That Can Affect the Net System Response

Internal to the Crossover Network

- Amplitude response of each filter.
- Phase response of each filter.
- Amount of overlap at each crossover frequency.
- Passband gain of each filter.
- Crossover/filter component quality.

External to the Crossover Network

- Impedance response of each driver.
- Acoustic response of each driver.
- Phase response of each driver.
- Relative sensitivity of each driver.
- Mounting location and alignment of each driver.
- Box type (closed, vented, bandpass, passive radiator).
- Box shape (diffraction effects).
- Series resistance of speaker cables and connectors.
- Amplifier source resistance.
- Acoustical environment of the listening space.

Because there are so many variables that affect the final response of a speaker, crossover network design usually involves considerable testing and tweaking. This is discussed in Chapter 7 ("Verifying & Adjusting the Design").

Crossover Network Topology

When a crossover network is designed and assembled its filters can be combined in several different ways, referred to as the network topology. They can be combined in parallel, series or a series-parallel mixture. The first two are shown below for a two-way crossover network:

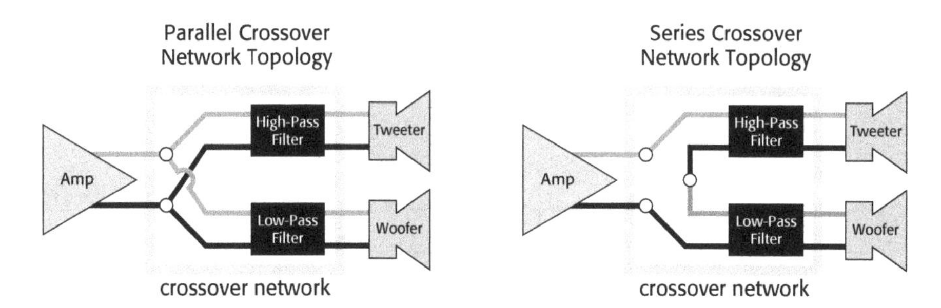

A parallel topology is widely considered to be the best because it allows each driver to be treated independently and each filter is less sensitive to the effects of the other filters in the network. The series or series-parallel topologies are more vulnerable to component variations than the parallel topology. Because the filters in a series or series-parallel network are interrelated the network is more complicated to design and must be designed as a whole. X•over Pro designs only crossover networks with parallel topologies.

Note: Some speakers use more than one of the same driver. For example, a speaker may have one tweeter, one midrange driver and <u>two</u> woofers. The woofers can be wired in parallel, series or (if four or more woofers are used) series-parallel. This is not the same as the network or filter topology. X•over Pro accommodates any of these multi-driver wiring methods and the relevant filter in the crossover network will view them as a single driver with the "net" parameters that result from their combination.

Filter Topology

The components of a filter can be arranged in numerous ways and this has a strong influence on the filter design. The way the components are arranged is the "topology" of the filter. X•over Pro uses the versatile "LC ladder" circuit topology for all low-pass, band-pass and high-pass filters. It is called an "LC ladder" topology because it contains inductors (L) and capacitors (C) and the circuit layout looks like a ladder that is laying on its side. A sample 4th-order bandpass filter is shown below:

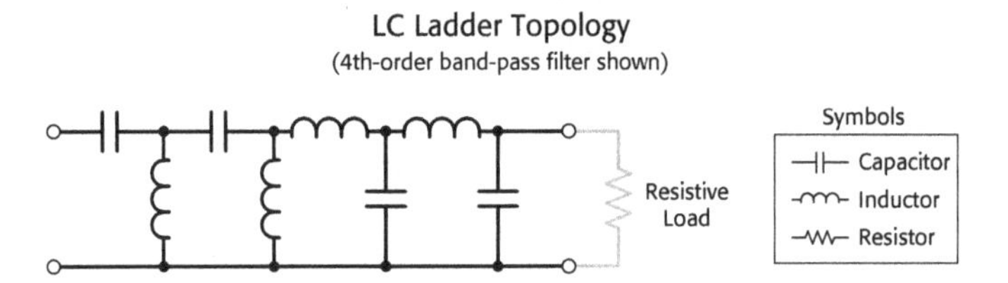

Using this topology a low-pass filter uses capacitors for the parallel rungs of the ladder and inductors for the series sides of the ladder. A high-pass filter is just the opposite. It uses inductors for the parallel rungs of the ladder and capacitors for the series sides of the ladder. A band-pass filter uses a combination of both since it is comprised of both a low-pass and high-pass circuit (shown above).

It is important that the "load" appear to be resistive or the filter will not perform as expected. This is a problem because a speaker driver is a reactive load with an impedance that can vary quite a lot at different frequencies. One solution is to equalize the impedance of the driver so that it looks mostly resistive to the filter. X•over Pro can help you do this because it has the capability of designing an impedance equalization circuit (see Chapter 3). Another solution is to build the network and experiment with its component values to determine what adjustments are necessary to make it work as intended with the selected driver.

It is also important that there be practically no input resistance to the filter. This is usually not a problem for modern amplifiers which have an extremely low output resistance and complementarily high damping factors.

Filter Order

Filters are classified several ways depending on the shape of their response. The first classification is called the "order" of the filter. The "order" is determined by counting the total number of capacitor and inductor sections in the filter and it describes how fast the filter will attenuate sound in the stopband. *Note: Because a band-pass filter is actually a combination of a low-pass and a high-pass circuit, its number of capacitor and inductor sections is divided by two to calculate its order.* The attenuation rate of the filter is measured in dB per octave and is called the "slope" or "cutoff rate" of the filter. The first four filter orders and their slopes are illustrated in the amplitude response graph below (see the "Normalized Amplitude Response" topic in Chapter 5 of the *X•over Pro Reference* section later in this manual for a detailed description of each part of a filter's response curve):

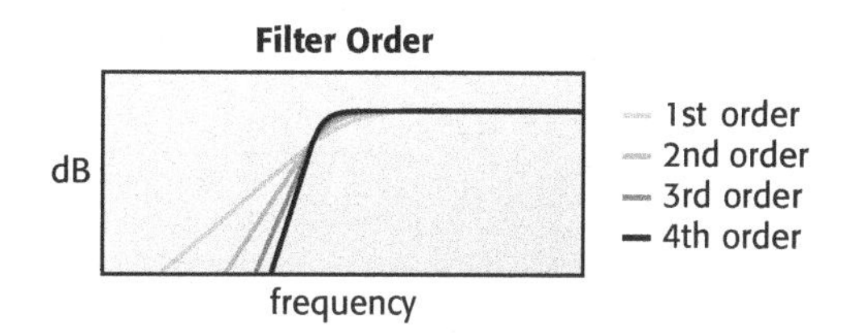

- 1st-order — filter slope is 6 dB/octave.
- 2nd-order — filter slope is 12 dB/octave.
- 3rd-order — filter slope is 18 dB/octave.
- 4th-order — filter slope is 24 dB/octave.

In general, filters with higher orders are less tolerant of component variations than filters with lower orders. *Note: Filters with a higher order than 4th-order are possible but they are rare and X•over Pro does not model them.*

Usually, the same order is selected for all filter sections in a crossover network. When this is the case, the crossover network can be said to have the same order. For example, a two-way crossover network containing a 4th-order low-pass filter and a 4th-order high-pass filter can be called a 4th-order two-way crossover network. However, the characteristics of the drivers may sometimes force you to select mixed order filters for a crossover network. X•over Pro allows you to independently select the order and type of each filter in a crossover network.

Choosing a filter order can be a complex task and depends on many things, including the internal and external factors mentioned previously (page 32). Sometimes the best way to select a filter is by experimentation. See the "Filter Summary" later in this chapter (pages 37-40) for further help.

1

Filter & Crossover Types

The filter type can be described in several different ways. Low-pass and high-pass filters in two-way crossover networks are often identified by their "Q". The Q is the resonance magnification of the filter and it is recognized by the shape of the "knee" of the amplitude response. Filters with a high Q tend to "ring" and exhibit poor transient response. Unlike drivers and boxes which use only numerical values for Q, filters are sometimes named after the engineer(s) who first described them. Some examples are shown in the amplitude response graph below (see the "Normalized Amplitude Response" topic in Chapter 5 of the *X•over Pro Reference* later in this manual for a detailed description of each part of a filter's response curve):

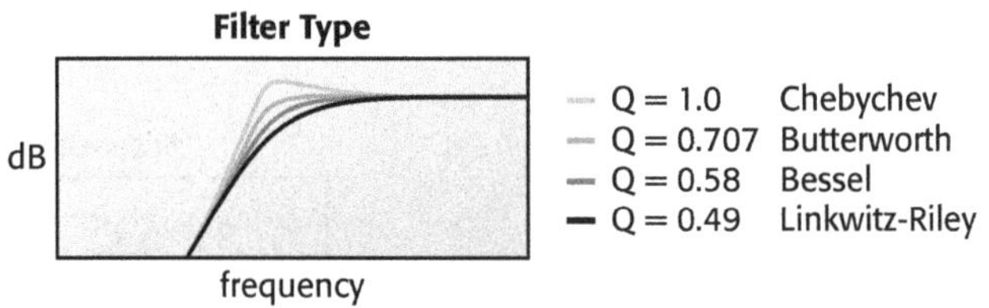

X•over Pro uses the common names, like Butterworth, to identify the filters in two-way crossover networks. See the "Filter Summary" topic on the next page for descriptions.

The filters in three-way crossover networks (and some two-way networks) are often identified as either "APC" or "CPC" depending on the way they combine. APC stands for "All-Pass Crossover" and it refers to those crossover networks whose filters sum to create a flat voltage output. APC networks are generally considered the best choice because they make it possible for the speaker to have a flat on-axis amplitude response. Common APC networks include 1st- and 3rd-order Butterworth filters and 2nd- and 4th-order Linkwitz-Riley filters. CPC stands for "Constant-Power Crossover" and it refers to those crossovers whose filters sum to provide a flat power response. The power response of a speaker is the total of both its off-axis and on-axis amplitude response. In other words, it is the total acoustical power that is radiated into a space. CPC networks can be beneficial in reverberant environments where the off-axis response is important.

The difference between APC and CPC networks can be understood electrically by a comparison of their input to output voltages. APC networks satisfy the following expression:

$$|V_I| = |V_L + V_M + V_H|$$

This means the absolute value of the input voltage will equal the absolute value of the sum of the output voltages of each filter at all frequencies. CPC networks satisfy the following:

$$V_I^2 = V_L^2 + V_M^2 + V_H^2$$

This means that the square of the input voltage will equal the sum of the squares of the output voltages of each filter at all frequencies.

Filter Summary

Each of the filter choices in X•over Pro are described next. These generalizations assume that the drivers are properly aligned at the crossover frequency. This means that they are mounted in such a way that the direct sound from each driver arrives at the listener's ear at the same time at the crossover frequency. Another important assumption is that the impedance response of each driver has been equalized so that it appears to be approximately resistive to the crossover network. (See Chapter 3 for information on impedance equalization.) Also, the sensitivity of the drivers is assumed to have been equalized with an appropriate L-pad. (See Chapter 4 for information on L-pad design.)

Finally, the following descriptions assume that all filters in the crossover network are of the same type. If a two-way crossover network has a 4th-order Linkwitz-Riley low-pass filter, it is assumed that it also has a 4th-order Linkwitz-Riley high-pass filter. Although X•over Pro allows you to mix different filter orders and filter types in the same crossover network, the results can vary widely and this "Filter Summary" makes no attempt to describe the behavior of various combinations. If you choose to use mismatched filters, you'll have to rely on the performance graphs and your own measurements and experience to determine the results.

1st-order Filters

Advantages: Can produce minimum phase response (Butterworth only) and a maximally flat amplitude response. Requires the fewest components.

Disadvantages: Its 6 dB/octave slope is often too shallow to prevent modulation distortion, especially at a tweeter's resonance frequency. Achieving minimum phase and a maximally flat amplitude response requires very careful driver alignment and only occurs when the listener is located at exactly the same distance from each driver. It has a 90° phase shift which can result in lobing and tilting of the coverage pattern.

Two-Way

1st-order Butterworth: Produces a –3 dB crossover point to achieve a maximally flat amplitude response, minimum phase response and flat power response that qualifies it as both an APC and CPC network. The 90° phase shift results in a –15° tilt in the vertical coverage pattern if the tweeter and woofer are vertically separated by no more than one wavelength at the crossover frequency and if the acoustical depth of the tweeter and woofer are carefully aligned at the crossover frequency. The tilt will increase and lobing can become severe if the drivers are separated by a greater distance or are misaligned. These problems appear as a ripple in the amplitude response. Filter Q = 0.707.

Two-Way & Three-Way

1st-order Solen Split –6 dB: A custom version of the 1st-order Butterworth filter (two-way crossovers) or 1st-order APC filter (three-way crossovers) that uses a –6 dB crossover point to minimize the disadvantages of a crossover network with standard 1st-order Butterworth or APC filters.

Three-Way

Note: 1st-order filters are usually not recommended for three-way crossover networks because their shallow 6 dB/octave slopes do not provide adequate separation.

1st-order APC: Produces –3 dB crossover points to achieve a flat amplitude response.

1st-order CPC: (Seldom used.) Produces –3 dB crossover points to achieve a flat power response.

2nd-order Filters

Advantages: Can produce a maximally flat amplitude response. Requires relatively few components. Has a 180° phase shift which can often be accommodated by reversing the polarity of the tweeter and which produces minimal or no lobing or tilt in the coverage pattern. Is less sensitive to driver misalignment than 1st-order filters.

Disadvantages: Although the 12 dB/octave slope is better than a 1st-order filter, it may still be too shallow to minimize the modulation distortion of many drivers.

Two-Way

2nd-order Bessel: Produces a –5 dB crossover point to achieve a nearly flat (+1 dB) amplitude response. The summed group delay is flat. It has a low sensitivity to driver misalignment and resonance peaks. Filter Q = 0.58.

2nd-order Butterworth: Produces a –3 dB crossover point that sums to a +3 dB amplitude response and a flat power response that qualifies it as a CPC network. It has a medium sensitivity to driver misalignment and resonance peaks. Filter Q = 0.707.

2nd-order Chebychev: (Seldom used.) Produces a 0 dB crossover point to achieve a +6 dB amplitude response with about ±2 dB of ripple. The summed group delay has a significant peak just below the crossover frequency. It has a medium sensitivity to driver misalignment and resonance peaks. Filter Q = 1.0.

2nd-order Linkwitz-Riley: (Very popular.) Produces a –6 dB crossover point to achieve a maximally flat amplitude response that qualifies it as an APC network. It has a –3 dB dip in the power response. The summed group delay is flat. It has a medium sensitivity to driver misalignment and resonance peaks. Filter Q = 0.49.

Three-Way

2nd-order APC: Produces –6 dB crossover points to achieve a flat amplitude response but the power response will have approximately 3 dB of ripple.

2nd-order CPC: (Seldom used.) Produces –3 dB crossover points to achieve a flat power response but the amplitude response will have approximately 3 dB of ripple.

3rd-order Filters

Advantages: Can produce nearly flat amplitude response. With an 18 dB/octave slope, it is better able to minimize modulation distortion. Less sensitive to driver misalignment.

Disadvantages: Requires more components. Has a 270° phase shift which can result in lobing and tilting of the coverage pattern.

Two-Way

3rd-order Butterworth: (Popular for some D'Appolito mid-tweeter-mid designs.) Produces a –3 dB crossover point to achieve a maximally flat amplitude response and flat power response that qualifies it as both an APC and CPC network. A 270° phase shift results in a +15° tilt in the vertical coverage pattern if the tweeter is wired with normal polarity and a –15° tilt if the tweeter is wired with reverse polarity. (D'Appolito mid-tweeter-mid designs overcome much of this tilt problem and produce a more symmetrical coverage pattern.) It has better group delay than a 1st- and 2nd-order Butterworth network. Filter Q = 0.707.

Three-Way

3rd-order APC: Produces –3 dB crossover points to achieve a flat amplitude response but the power response will have a modest ripple (usually less then 1 dB) that increases slowly as the spread between the two crossover frequencies increases.

3rd-order CPC: (Seldom used.) Produces –3 dB crossover points to achieve a flat power response but the amplitude response will have a varying amount of ripple (typically 1 to 3 dB) depending on the spread between the two crossover frequencies.

4th-order Filters

Advantages: Can produce a maximally flat amplitude response. With a 24 dB/octave slope it provides the best isolation between drivers resulting in the least modulation distortion. Has a 360° phase shift which results in "in-phase" response and which promotes minimal or no lobing or tilt in the coverage pattern. Is the least sensitive to driver misalignment.

Disadvantages: Requires the most components. The increased number of inductors can result in substantial insertion loss because of inductor DCR.

Two-Way

4th-order Bessel: Produces a –5 dB crossover point to achieve a nearly flat (+1 dB) amplitude response. The summed group delay is flat. Filter Q = 0.58.

4th-order Butterworth: Produces a –3 dB crossover point that sums to a +3 dB amplitude response and flat power response that qualifies it as a CPC network. The summed group delay has a significant peak just below the crossover frequency. Filter Q = 0.707.

1

4th-order Gaussian: (A seldom used filter that is constructed with an asymmetrical filter topology.) Produces a –6 dB crossover point to achieve a nearly flat amplitude response with moderate ripple. The summed group delay produces a moderate bump just below the crossover frequency.

4th-order Legendre: (A seldom used filter that is constructed with an asymmetrical filter topology.) Produces a –1 dB crossover point that sums to a +5 dB amplitude response with minor ripple. The summed group delay has a significant peak just below the crossover frequency.

4th-order Linear-Phase: (A seldom used filter that is constructed with an asymmetrical filter topology.) Produces a –6 dB crossover point to achieve a nearly flat amplitude response with moderate ripple. The summed group delay produces a moderate bump just below the crossover frequency.

4th-order Linkwitz-Riley: (Very popular. Sometimes called a "squared Butterworth" filter. Also used for some D'Appolito mid-tweeter-mid designs.) Produces a –6 dB crossover point to achieve a maximally flat amplitude response that qualifies it as an APC network. It has a –3 dB dip in the power response. The summed group delay produces a moderate bump just below the crossover frequency. Filter Q = 0.49.

Three-Way

4th-order APC: Produces –6 dB crossover points to achieve a flat amplitude response but the power response will have approximately 3 dB of ripple.

4th-order CPC: (Seldom used.) Produces –3 dB crossover points to achieve a flat power response but the amplitude response will have approximately 3 dB of ripple.

Note: If some of the preceding terminology is unfamiliar, please consult the "Glossary of Terms" in Appendix B near the end of the manual.

2 Open & Sealed Back Drivers

Although piston drivers come in many different sizes and shapes, they can usually be sorted into two groups: drivers with open backs and drivers with sealed backs (shown below).

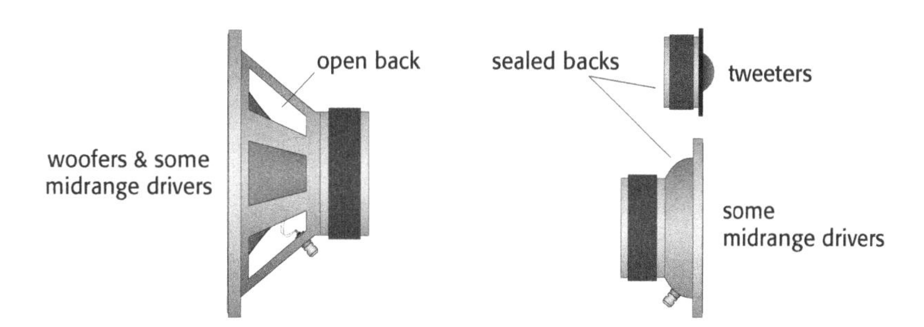

Sealed back drivers create sound waves on the front side only. They are not affected by the compliance of the air inside the speaker box. X•over Pro assumes that all tweeters have a sealed back.

Open back drivers expose both the front and back side of the diaphragm so they create sound waves on both sides. However, the sound waves which emanate from the back side have the opposite polarity as those which emanate from the front. Because of this, open back drivers require a box to trap the rear sound waves so that the longer, low-frequency ones do not mix with and cancel the front ones. This is illustrated below. *Note: High frequency sound waves are directional and do not mix between front and back–even when no box is present.*

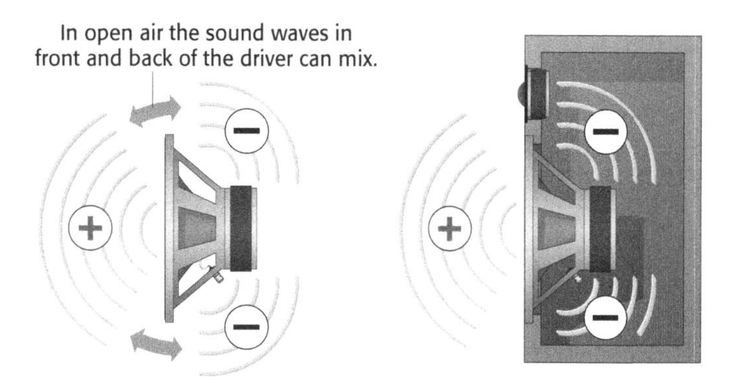

Open back drivers are "loaded" by the air inside the speaker box, controlling their low-frequency response. X•over Pro assumes that all woofers have an open back.

Midrange drivers fall into both groups—some have open backs and some have sealed backs. This is specified in X•over Pro by the "Midrange Type" selection on the "Description" tab of the Driver Properties window (see the "Description" topic of Chapter 2 in the *X•over Pro Reference* later in this manual).

2

When a Box is Required

The moving coil electrodynamic piston driver model in X•over Pro expects all open back drivers to have a box. Box designs and driver parameters can be imported from a BassBox Pro or BassBox Lite design file. If no box information is provided, X•over Pro assumes that an open back driver is mounted in an infinitely large closed box (infinite baffle).

Although X•over Pro is not a box design program, it does allow the user to enter basic box information on the "Impedance EQ" tab of the Driver Properties window (see the "Impedance EQ" topic of Chapter 2 in the *X•over Pro Reference*). Why the "Impedance EQ" tab? A speaker box has a strong influence on the impedance response of an open back driver and X•over Pro provides the ability to compensate for the system impedance of the driver in the box with an impedance equalization network. This is why the box parameters are located on the "Impedance EQ" tab. If desired, see Chapter 3 for more information about impedance equalization.

Also note that the effects of the box (or infinite baffle if no box information is available) is included in all graphs for open back drivers.

3 Driver Impedance Equalization

Each filter in a crossover network would like the driver or drivers attached to it to be purely resistive. This means that they would have a flat impedance response. For example, an 8 ohm driver should have an impedance of 8 ohms at all frequencies. Unfortunately this is never the case. Two typical impedance response curves are shown below:

Impedance Response

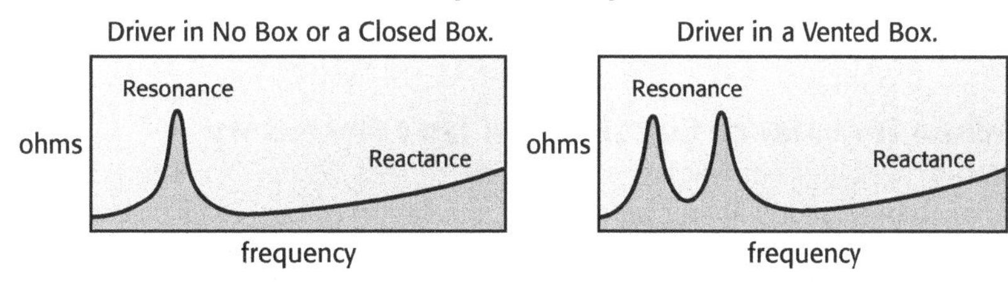

There are two characteristics that dominate the shape of the impedance response: 1) one or more resonance peaks near the lower limit of the driver's response and, 2) an inductive reactance rise at the upper limit of the driver's response.

Resonance

All drivers have a resonance. This often produces a sharp impedance peak like the one shown in the left graph above. The magnitude of the impedance resonance peak is moderated in some drivers. A common example is tweeters with ferrofluid. The ferrofluid has the benefit of greatly reducing the height of the resonance peak.

The box can also affect the resonance. A closed box will usually change the frequency and magnitude of the resonance, moving the location and height of the impedance peak. A vented box does this and more. It adds a second impedance peak because it has a vent with a second resonance. This is shown in the right graph above.

If the crossover frequency or frequencies of the network can be kept at least one octave (preferably two or more octaves) away from these impedance peaks, they should not pose a problem for the performance of the network.

Reactance

Most drivers also have reactance. This is especially true of drivers with voice coils–the vast majority. Because a voice coil is a coil, it's primary reactance is inductive. This inductance is represented by the driver parameter "Le". Inductive reactance acts like a slow low-pass filter and causes the impedance to gradually rise as the frequency increases.

Is the reactance always a problem? Not if it is low enough so that the region on either side of each crossover frequency is flat. Ideally the crossover network would like the impedance to be flat for at least one octave (preferably two or more octaves) on either side of each crossover frequency.

Fortunately there is a solution for non-flat driver impedance. The impedance can be flattened with an impedance equalization network (also referred to as a "Zobel" network). This is a circuit that is placed between the crossover network and the driver.

An impedance equalization network will equalize the impedance of just one load. A three-way crossover network with a low-pass, band-pass and high-pass filter may need as many as three different impedance equalization networks, one for the output of each filter.

When Impedance Equalization Isn't Necessary

It was mentioned earlier that impedance equalization is not needed if the impedance response of a load is flat for at least one octave on either side of a crossover frequency. Are there other times when impedance equalization can or should be avoided? Yes.

Impedance equalization of one or more resonance peaks is usually not desired when the crossover frequency is below 300 Hz. This is because large capacitors and inductors are required to equalize low frequency resonance peaks and this can introduce other problems like substantial insertion loss. However, this problem does not exist for impedance equalization networks that equalize only the inductive reactance rise of a voice coil. An impedance equalization network can still be used as long as resonance peak equalization is avoided. *Note: When low frequency resonance peaks are present and you can't equalize them, you can still build a good crossover network. It will just require more experimentation.*

Impedance Equalization Network Types

X•over Pro provides the four different impedance equalization circuits shown below.

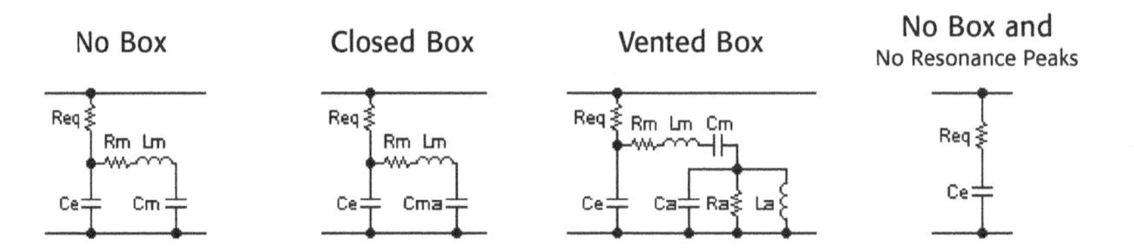

They are selected with the "Impedance EQ" tab of the Driver Properties window (see the "Impedance EQ" topic of Chapter 2 in the *X•over Pro Reference* section later in this manual). This is also the place where the box information is entered since it affects the impedance response.

4 Matching Driver Levels

For high fidelity, a speaker should have a flat amplitude response. This means that the sound waves of each driver need to sum to a flat response as shown below:

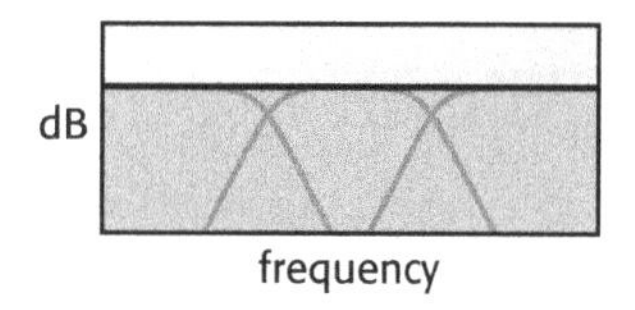

This assumes that each driver in a multi-way speaker has the same efficiency. What happens if one of the drivers is more efficient than the others? The illustration below depicts a three-way speaker with a more efficient tweeter:

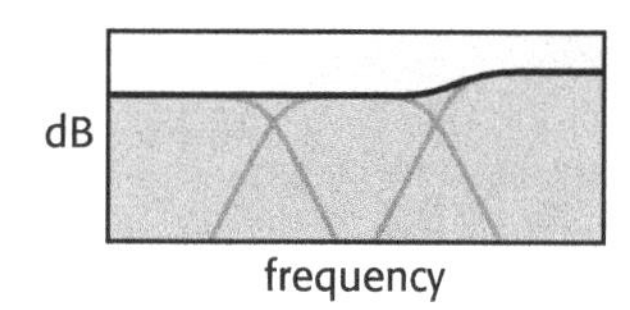

Unfortunately, drivers seldom have the same efficiency. Many woofers are less efficient than comparable midrange drivers. And many midrange drivers are less efficient than comparable tweeters.

There are several ways to deal with this problem. One way is to double up on the less efficient drivers. For example, use two woofers instead of one. This can give you as much as a 6 dB increase in loudness for the "bottom end" and help bring it up to the level of the mids and highs. (The amount of increase depends on the electrical and mechanical configuration. For more information see the "Configuration" topic of Chapter 2 of the *X•over Pro Reference* later in this manual.)

Another way to deal with the problem of mismatched sensitivities is to attenuate the louder drivers. This is less expensive than adding more drivers but it does require that one or more attenuator circuits be added to the network. The most common type is an "L-pad" composed of two resistors as shown below. It gets its name from its "L" shape.

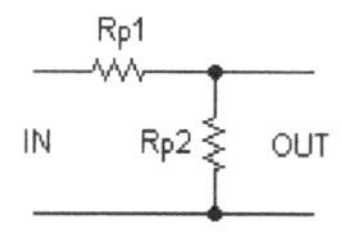

The L-pad circuit should be located between the crossover network and the driver. If an impedance equalization circuit is also used, the L-pad should be located between the crossover network and the impedance equalization network.

In X•over Pro L-pads are designed with the L-pad Attenuator window. It is opened with the "L-pad" command in the Network menu of the main window or with the keyboard shortcut Ctrl+A. For more information, see Chapter 4 of the *X•over Pro Reference*.

5 Managing Low Band-Pass Impedance

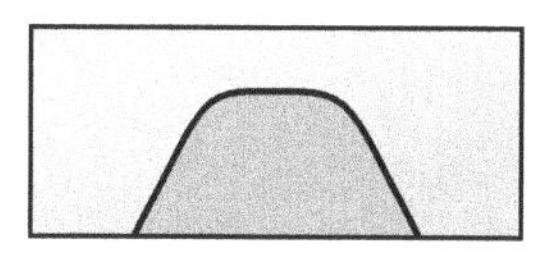

Designing a band-pass filter in a three-way crossover network can sometimes be very challenging. The band-pass filter design must handle the complex interaction of its internal high-pass and low-pass filter sections. The APC and CPC band-pass filters calculated by X•over Pro do a very good job and produce consistent, predictable results as long as the midrange driver has a flat impedance response for at least one octave on either side of the passband. However, the band-pass filters in X•over Pro can easily result in an impedance that is less than half that of the net nominal impedance of the midrange driver(s). For example, the amplifier may see an impedance in the midrange passband that is less than 4 ohms if the midrange driver has a nominal impedance of 8 ohms. If the midrange driver has a 4 ohm impedance, then the network may be less than 2 ohms in the midrange passband.

Is this a problem? If the impedance does not drop below the rated minimum load impedance of your amplifier, then you should not have a problem. Even if it does, it may still be okay because the impedance will only be low in the passband of the band-pass filter. It should not be as low in the low-pass filter that feeds the woofer so the demands on the amplifier output will probably not be excessive. Please consult the manufacturer of your amplifier for further details about its load handling capabilities at different frequencies.

Can the impedance of the band-pass filter be increased? Yes—if the midrange driver has a greater sensitivity than the woofer and tweeter. A procedure is listed below along with some points to consider.

- **Choose a midrange driver that has a greater sensitivity than the woofer and tweeter.** This is necessary because raising the passband impedance of the band-pass filter and midrange driver(s) will reduce their sensitivity. This will be accomplished with optional resistor R1 in the band-pass circuit. (R1 is in series with the midrange driver and is provided to reduce the passband gain common to many band-pass filters.)

  An example of a 3-way crossover network is shown at right. It includes all 4th-order filters. Notice that resistor R1 is highlighted with a light grey circle in the band-pass circuit between the last inductor (L6) and the midrange driver.

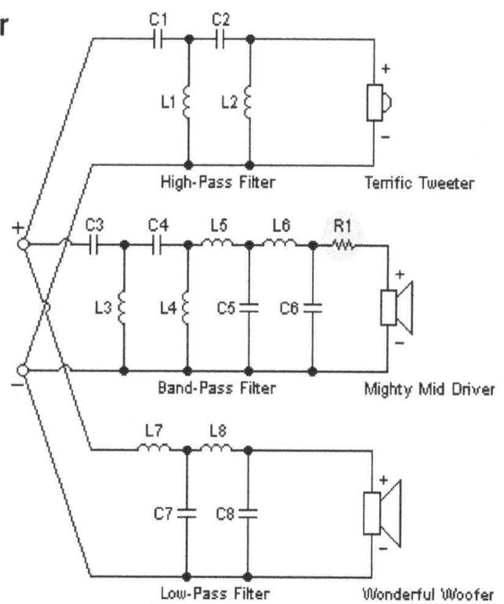

- **Use R1 to raise the impedance of the band-pass filter.** Here is the procedure:
  1 Enter the driver parameters including the Re and Z (pages 104-112). *Note: It is strongly recommended that sufficient parameters be entered for each driver so that their response can be displayed in the graphs because the graphs are an important part of this procedure.*
  2 If appropriate, design an impedance equalization network for the relevant drivers (pages 120-127).
  3 Select "3-way Crossover" from the Network menu. Make sure that the "Omit" check-box at the bottom of the "Filters" tab of the main window is not checked. Enter the filter frequencies, select the filter types and design the crossover network. X•over Pro will calculate the component values for the crossover network–including resistor R1.
  4 Record the value of R1 and then begin to manually increase it until the level of the midrange driver is approximately equal to the woofer and tweeter. You will probably see ripple in the band-pass response as you begin to increase the value of R1. You can disregard the ripple for now. Record how much R1 was increased.
  5 Open the Driver Properties window and temporarily increase the value of the midrange driver's Re value the same amount that you increased R1. If the midrange driver has an L-pad, open the L-Pad Attenuator window and increase the value of the midrange driver's Zt parameter the same amount that you increased R1. Then select "Recalculate Now!" (Ctrl+R) from the Network menu. This will cause X•over Pro to recalculate the band-pass filter and adjust it for the increase of R1.
  6 Restore the midrange driver Re parameter back to its original value. If a midrange driver L-pad is present, restore its Zt and Attenuation parameters back to their original values. Add the additional resistance back to R1 (it was reset when the crossover network was recalculated in Step 5). If needed, you can return to Step 4 and repeat the last part of this procedure to tweak the value of R1 until the level of the midrange driver matches that of the woofer and tweeter.
  7 If the midrange driver has an L-pad, reduce its Attenuation level an appropriate amount (or remove it altogether) so that the amplitude response of the midrange driver is restored to an appropriate level.
- **Avoid band-pass filters with a narrow passband.** It is best to have at least 3 octaves between the lower and upper corner frequencies.
- **Choose a midrange driver with a DC resistance (Re) of 6 ohms or greater.** *Note: Re has a much greater effect on the impedance of the band-pass filter than the driver impedance (Z). Re will usually be 80% of Z.*

Finally, if the desired impedance cannot be obtained with a passive three-way crossover network, consider using an active crossover network instead of a passive crossover network.

5

Sample

Let's examine a sample to see how the preceding seven steps are applied. We will use a three-way crossover network design that was included with X•over Pro. To view it, open file "low BP impedance sample.xo3" in your "Designs" folder. If you used the default settings when you installed X•over Pro, the path will be: "c:\Program Files\HT Audio\Designs".

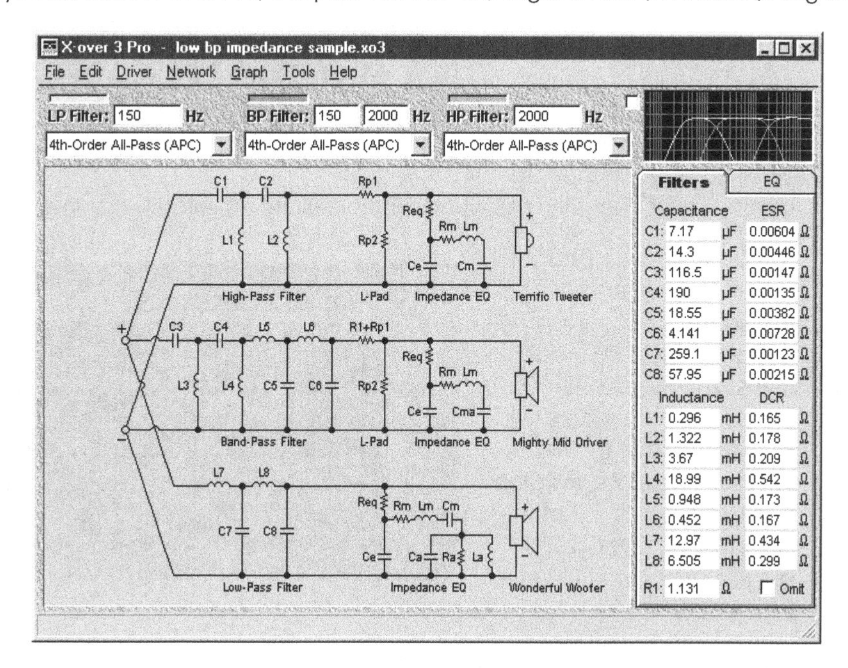

Note: This sample assumes that Re (rather than Z) is used as the basis for all calculations. Check the "General" tab of the Preferences window (page 211) to make sure that the "Base calculations on" setting is set to "Driver DC Resistance (Re)" as shown at right:

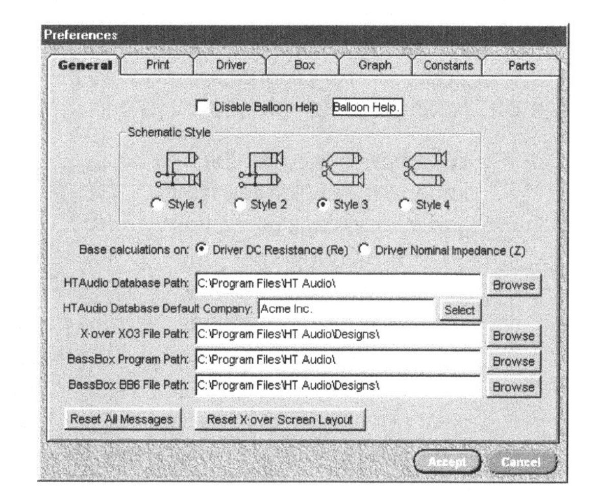

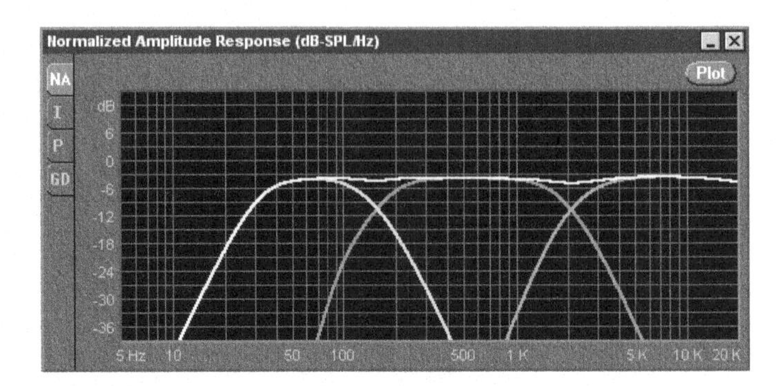

This sample contains fictitious drivers in a 4th-order APC crossover network. Each driver has a nominal impedance (Z) of 8 ohms and each driver has an impedance equalization network. The midrange driver includes a 3.5 dB L-pad and the tweeter includes a 2.5 dB L-pad because they are both more efficient than the woofer. The resultant amplitude response is shown above:

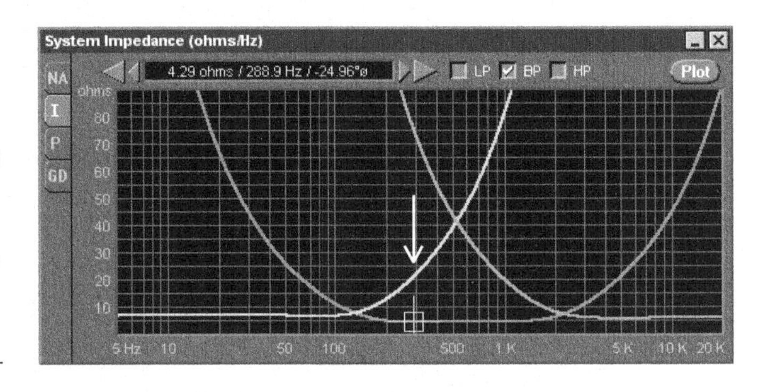

The DC resistance (Re) values of each driver vary. The woofer Re = 6.5 ohms. The midrange driver Re = 5.2 ohms. The tweeter Re = 5.9 ohms. These Re values lead us to expect that the net system impedance will drop at least as low as 5.2 ohms and probably lower. Let's see just how low in the illustration at right:

Notice the we turned off the net plot line in the graph above. This makes it easier to focus on the impedance of the band-pass filter. Using the cursor on the band-pass (BP) curve we see that the band-pass filter drops to a minimum impedance of just 4.29 ohms centered at 288.9 Hz. For comparison, the low-pass filter has a low impedance of 6.52 ohms at 81.31 Hz and the high-pass filter has a low impedance of 5.67 ohms at 4.697 kHz.

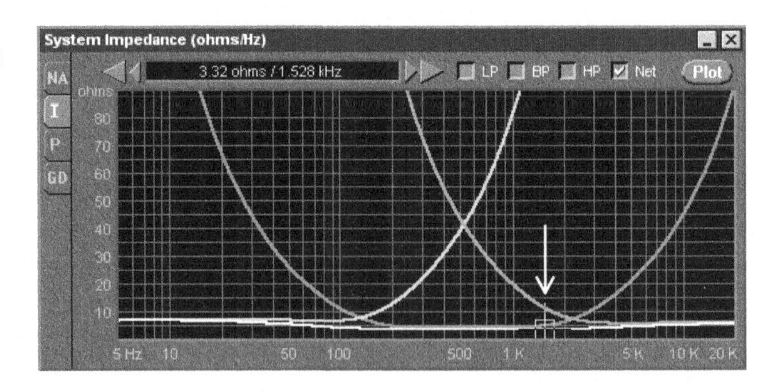

What happens when the impedance of all three filters and their associated drivers are summed together? The graph at right includes the net plot line to answer this question. Remember that the net plot line shows us what our amplifier will "see".

5

The lowest net impedance is 3.32 ohms at 1.528 kHz (marked with the arrow above). A second low impedance point of 3.33 ohms is located at 232.4 Hz. These low points may be trouble to our amplifier which has a low-impedance rating of only 4 ohms. *Note: Most 4-ohm rated amplifiers can probably drive a 3.32 ohm load with no problem. But we will assume that our amplifier cannot for the purpose of this exercise.*

Our goal will be to raise the net impedance of the band-pass filter / midrange driver so that the net impedance of the speaker never drops below 4 ohms. So far in this sample, the first three steps have been completed because we started with an existing design. The actual Z and Re of all drivers have been entered (**Step 1**); an impedance equalization network has been designed for each driver (**Step 2**); and the crossover network has been designed and its component values calculated, including R1 (**Step 3**). Lets continue with Step 4:

Step 4 – We made a note that the value of the band-pass gain series resistor, R1, had a beginning level of 1.131 ohms. Next, we increased its value in order to increase the band-pass impedance. We decided to raise it just above the low-pass filter level of 6.52 ohms which we had observed earlier. To accomplish this we had to switch back and forth between adjusting R1 at the bottom of the "Filters" tab of the main window and replotting the System Impedance graph while observing the results. We found it helpful to temporarily turn off the "net" impedance plot line of the graph while doing this so the band-pass curve could be viewed more easily.

After experimenting for a while we decided to use a value of 5 ohms for R1. This represented an increase of 3.869 ohms (5 ohms = 1.131 + 3.869 ohms). After entering 5 ohms into the R1 field, we replotted the System Impedance graph as shown at right:

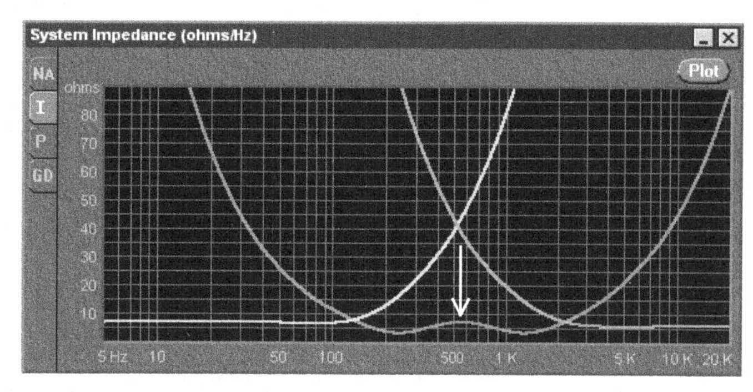

Ignoring the ripple to either side of the band-pass filter's center, we concluded that the additional 3.869 ohms looked good because it raised the impedance to 7.07 ohms at 554.4 Hz. This is about half an ohm higher than the low-pass filter's minimum impedance and should result in a net impedance that is higher than 4 ohms.

Step 5 – Next, we opened the Driver Properties window, selected the midrange driver in the title bar, selected the "Parameters" tab and raised the value of the midrange driver's Re by the same amount. Since the midrange driver's Re = 5.2 ohms, we changed it to 9.069 ohms (5.2 + 3.869). This is shown on the next page.

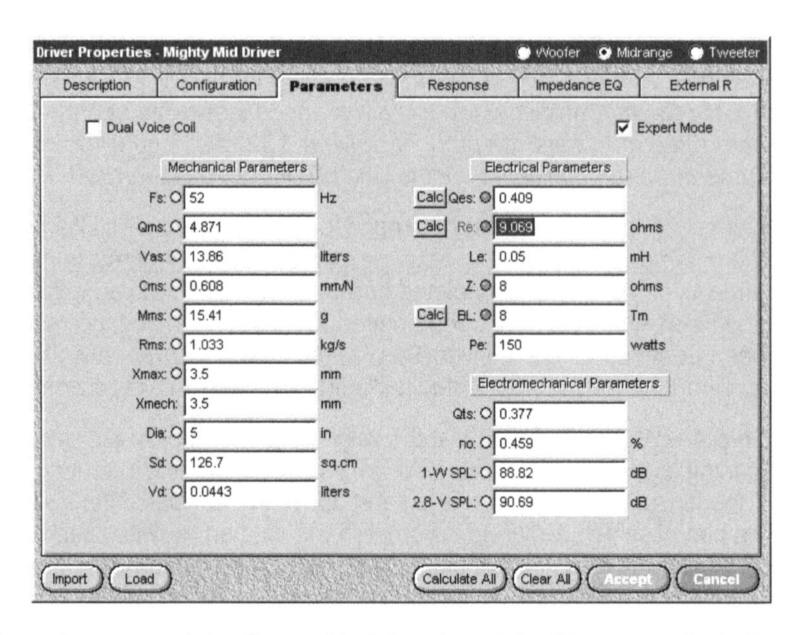

Notice above that we used the "Expert Mode" option of the "Parameters" tab. It analyzed the new value of Re and realized that something was wrong with the parameters so it turned several status indicators red. We ignored the status indicators because the original value of Re would be restored to 5.2 ohms in the next step.

This same adjustment also needs to be made to the midrange driver's Zt (total impedance) parameter of the L-Pad Attenuator window as shown at right:

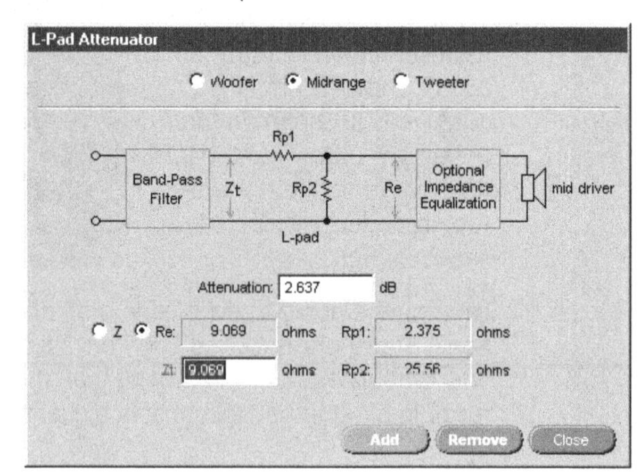

Notice that the 9.069 ohms now appears in the "Re" value of the L-Pad Attenuator window. Click in the Zt box and its background will turn white to show that it is ready to accept input. Enter 9.069 ohms so that Zt matches Re. *Note: Ignore the Attenuation, Rp1 and Rp2 values for the moment. We'll return to the L-pad in the next step.*

Remember to click the L-Pad Attenuator window's "Add" button after changing Zt so that the new value will be entered into the L-pad design. With both Re and Zt set to 9.069 ohms, the band-pass filter will now see a load impedance of 9.069 ohms.

Next, we used the "Recalculate Now!" command of the Network menu ([Ctrl]+[R]) to force X•over Pro to recalculate the band-pass filter component values using the new Re value of the midrange driver. We noted that resistor R1 equaled 1.972 ohms after the recalculation.

Step 6 – The last step was to remove the 3.869 ohm increase from Re and Zt and add it back to R1. We returned to the Driver Properties window and changed the Re of the midrange driver back to the original value of 5.2 ohms. Once again, the status indicators were all green. Then we closed the Driver Properties window with the "Accept" button. Afterward, we returned to the L-Pad Attenuator window and changed the Zt of the midrange driver back to 5.2 ohms. We also restored the Attenuation setting back to 3.5 dB which in turn restored the Rp1 and Rp2 values. Then we clicked on the "Add" button to enter the L-pad changes into the design.

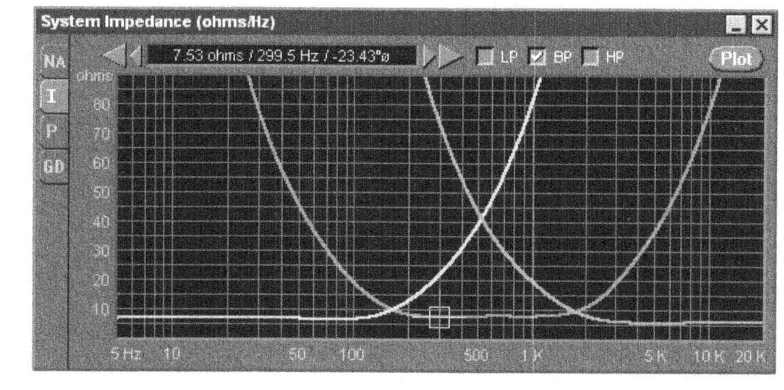

Next we added 3.869 ohms to R1 and entered a value of 5.841 ohms (1.972 + 3.869). Finally, we returned to the System Impedance graph and observed the results shown at right:

Notice that the minimum impedance of the band-pass filter is now 7.53 ohms at 299.5 Hz and that the response is once again smooth. The next graph shows the minimum net impedance:

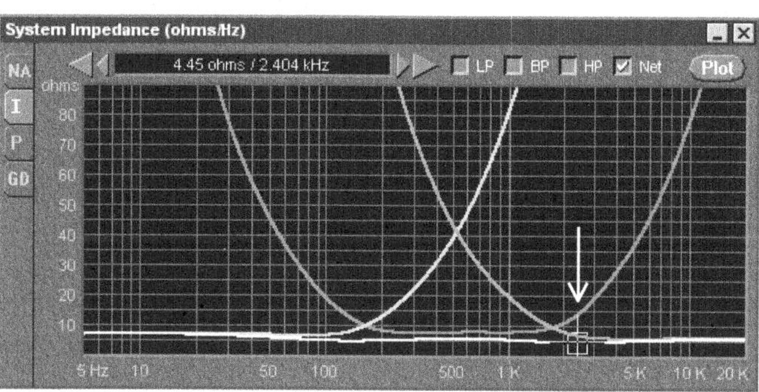

The minimum net impedance is now 4.45 ohms at 2.404 kHz with a second low point of 4.72 ohms at 132.6 Hz. Our goal was accomplished and our amplifier should now have no problem driving the speaker. But are we finished?

Step 7 – Let's return to the Normalized Amplitude Response graph and see how the amplitude response has changed.

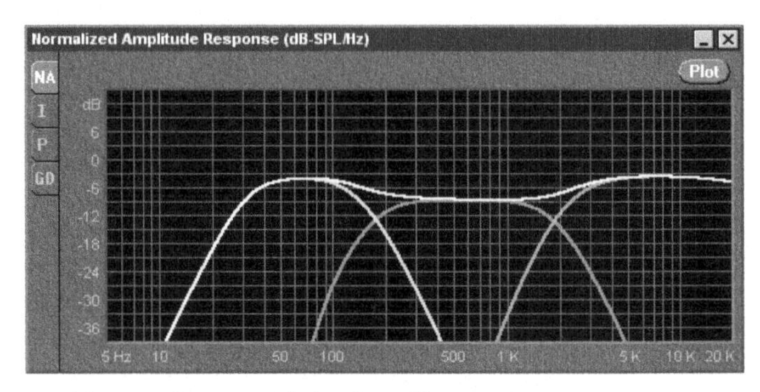

As shown at right, the level of the midrange driver is now too low. This is a disadvantage of increasing the resistance value of R1. It reduced the output of the midrange driver. Is there a solution? In this case, yes.

Remember that the midrange driver has a 3.5 dB L-pad because the midrange driver is more efficient than the woofer. Let's remove the L-pad and replot the Normalized Amplitude Response graph. The result is shown at right:

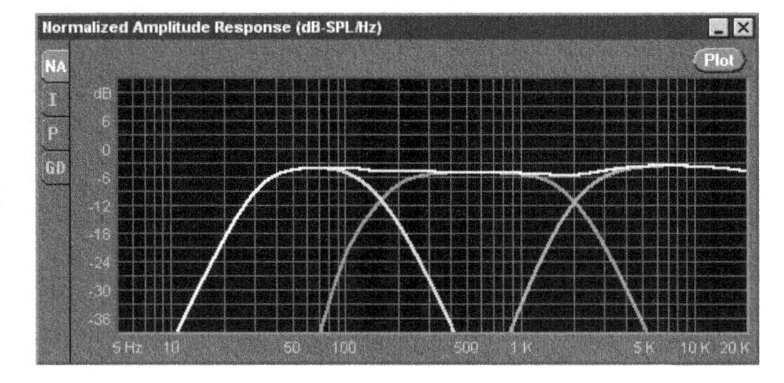

The net amplitude response is once again level. This underscores the importance of using a midrange driver with a higher sensitivity so that it can accommodate the loss in level that will occur when the band-pass impedance is raised.

This concludes the band-pass low-impedance example. What if the impedance of the low-pass or high-pass filter had been too low instead of the band-pass filter? In such a case we may have had to select a different woofer and/or tweeter with a higher Z and Re. If multiple drivers were used—for example, two woofers—then we could wire the woofers in series to raise their net impedance. We could also add a series resistor between a filter and driver. (This can be modeled in X•over Pro with series resistor Rp1 of the L-pad circuit. It is okay to leave the L-pad's parallel resistor, Rp2, with no value. As long as Rp2 equals zero or has no value, it will be ignored in the calculations even though it still appears in the schematics.) Lastly, we could switch to an amplifier with sufficient current headroom to drive the low impedance of our speaker. These are the trade-offs that must be considered when the avoidance of a low impedance is deemed an important element of speaker and crossover network design.

6 Crossover Network Construction

The purpose of the crossover network is to divide the sound between the drivers in a multi-driver speaker. How well this is accomplished is not just a matter of good circuit design. It also requires good execution. This chapter addresses network construction with such topics as component quality, mounting and wiring.

General

Let's begin with some general suggestions for the overall network. Later we will focus on individual component types.

Measure all components

It is best to measure the value of all components before using them in a crossover network. This is best accomplished with an impedance bridge and tone generator. Ideally, the components should be measured at the crossover frequency.

Lower-order networks like 1st and 2nd-order networks can usually tolerate greater component variations than higher-order networks like 3rd and 4th-order networks. If you are unsure whether a component value is close enough to work, try substituting the measured value in your X•over Pro design and plot the results. Use the graphs to evaluate the changes.

Series components are the most critical

The series components are the most important elements in a crossover network because the audio signal must pass through them before arriving at the driver. The parallel

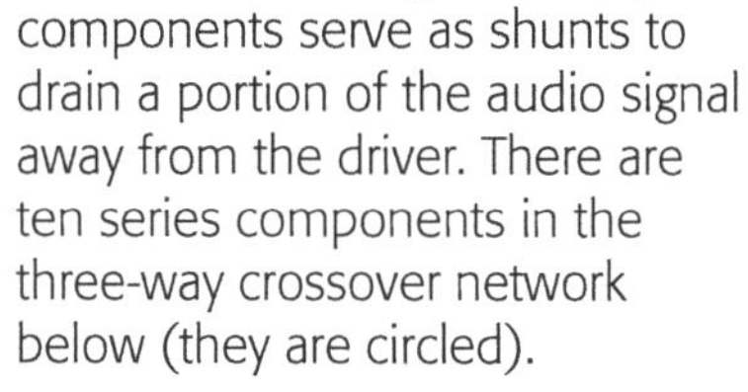

components serve as shunts to drain a portion of the audio signal away from the driver. There are ten series components in the three-way crossover network below (they are circled).

If you need to economize, it is best to do so with the parallel components and reserve the highest quality parts for the series components.

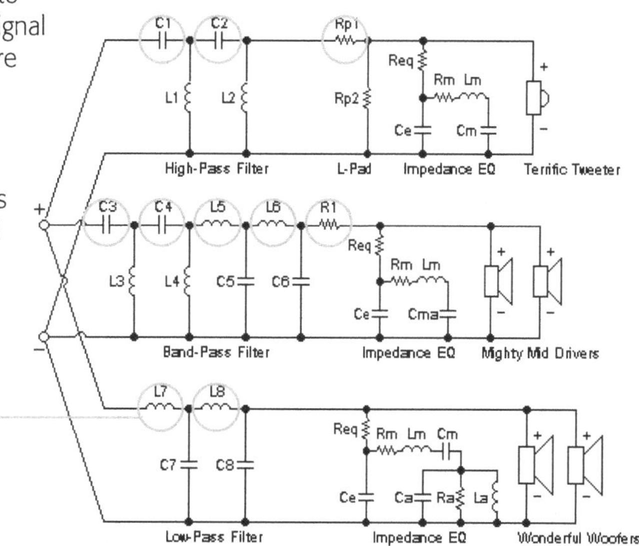

Connections

To make a good electrical connection between two or more components, begin by creating a strong mechanical connection. This can be as simple as carefully twisting their wire leads together. Take care not to put too much tension on each lead where it enters the component. A component with a broken lead is useless. Finally, solder the connections to create a long lasting, airtight connection. The wires should be heated enough to allow the solder to flow onto them but not so hot that a component is damaged.

Mounting

Unless you have the resources to design and manufacture a printed circuit board (PCB) for the crossover network you will need to obtain some sort of flat and thin nonconductive material to serve as a circuit board for component mounting. Commonly used materials include hardboard and pegboard.

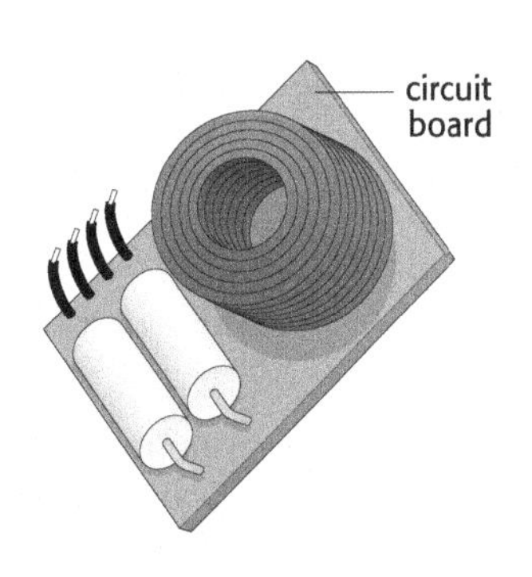

Create a cut pattern for the circuit board by laying all the components on a piece of paper in their mounting position and drawing a rectangle around them. Use the pattern to cut the correct size piece from the circuit board material.

When mounting the parts, it is important to not allow them to touch each other (except at the connections). This is especially true of inductors and resistors.

The parts should be securely mounted to the circuit board so that they cannot vibrate. This can be accomplished by gluing them to the circuit board with a nonconductive adhesive. Commonly used adhesives include silicone rubber and epoxy glue.

Most crossover networks are installed inside the speaker box. If this is done, it is important to keep the crossover network as far as possible away from the drivers. This is because the magnetic field of the drivers can interact with the crossover network and cause distortion. This problem is minimized when shielded drivers are used.

Some designers recommend using brass screws to mount the circuit board. Unlike steel screws, brass screws will not interact with the magnetic fields of the crossover network components.

Capacitors

Generally speaking there are two kinds of capacitors used in crossover network construction: electrolytic and solid dielectric. Electrolytic capacitors must be the nonpolarized variety because audio signals have alternating current. Electrolytics are the most common and the cheapest capacitors. Unfortunately they are considered the least favorable because they tend to have more resistance (ESR) and inductance (both undesirable qualities in capacitors) and because they do not age well. Electrolytic capacitors usually have greater manufacturing variations, resulting in higher tolerance values. And their resistive and inductive attributes are usually worse at higher frequencies.

Solid dielectric capacitors are usually considered much better choices because they have less resistance and inductance and are more stable as they age. However, they are also more expensive and are not usually available in the larger values. Examples of solid dielectric capacitors include polypropylene, polyester and Mylar.

When large capacitance values are needed, high-voltage oil-filled (motor-run) capacitors are sometimes used.

Use high quality capacitors in critical areas
The most important capacitors in a crossover network are the ones that are wired in series with the drivers. Whenever possible use higher quality capacitors for these locations. It is also a good idea to avoid the use of electrolytic capacitors for high-pass filters because they do not pass high frequency signals as well as solid dielectric capacitors. However there are solutions for this problem discussed below.

The most forgiving location for cheap capacitors is in the low-pass filter where they are used at lower frequencies and are wired in parallel with the driver to serve as shunts.

Improving capacitor performance
Some of the resistance and inductance problems of cheaper capacitors can be overcome with the addition of a high-quality bypass capacitor. These are usually small 0.1 to 2.0 µF Mylar capacitors that are paralleled with the cheaper capacitor. This can improve the high-frequency transient response considerably. Because the bypass capacitor is small it does not add much to the cost.

Another way to improve the sound quality of cheaper capacitors is to parallel several smaller ones in place of a single larger one. For example, use three parallel 50 µF capacitors in place of a single 150 µF capacitor. By paralleling them, the capacitance adds, but the bad resistive and inductive qualities become smaller. The corollary to this is to avoid wiring two or more cheaper capacitors in series with each other because this increases their undesirable qualities.

Inductors

Generally speaking there are two kinds of inductors used in crossover network construction: inductors with an air core and inductors with a metal core. Air core inductors are usually considered the best because they do not become saturated as quickly at higher power levels as comparable metal core inductors. This results in lower distortion and greater power handling for air core inductors. And, with a little patience, they can be made at home.

Metal core inductors offer the advantage of higher inductance with lower resistance (DCR) compared to comparable air core inductors. This is because the metal core increases the amount of inductance per turn of wire, causing them to need fewer turns for a given inductance value. This also makes them smaller than air core inductors—another advantage.

Mounting
Inductors generate their own magnetic fields and they are susceptible to the effects of external magnetic fields. To minimize magnetic coupling between inductors, it is generally recommended that they be mounted at 90° angles to each other and that they be at least 3 inches (76 mm) apart. This is shown below:

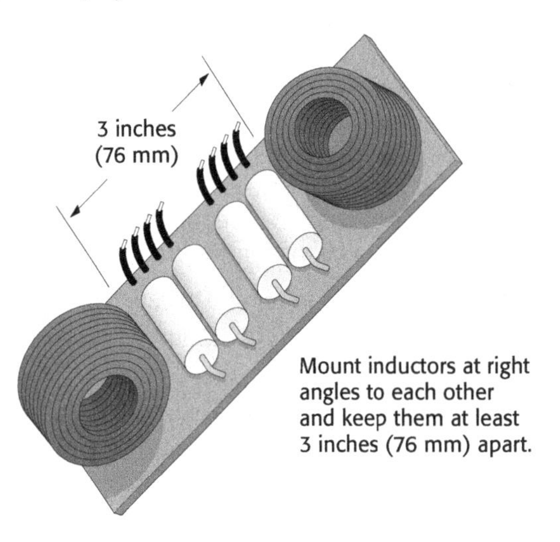

To minimize interference from external magnetic fields, they should not be located close to drivers, whose magnets can generate strong external magnetic fields (unless they are shielded).

Keep resistance low
The number one problem of inductors is excessive resistance. As a general rule, their DC resistance (DCR) should be no higher than 5% (1/20th) of the driver impedance (Z). This means that an inductor that is connected to an 8 ohm driver should have a DCR no higher than 0.4 ohms.

To keep the DCR low, you can use a metal core inductor. However, if you can tolerate the larger size, a better solution might be to use an air core inductor that is made with a larger gauge of wire. The heavier the wire, the better. 16 gauge is a good starting point for low-pass inductors and 18 gauge is a good starting point for midrange drivers and tweeters. *Note: The smaller the gauge number, the larger the wire and the smaller the resistance.*

Homemade inductors

If you wind your own air core inductors, it is important to wind them tightly and to secure the windings so they cannot vibrate. Some claim that vibrating windings will create audible distortion. The windings can be secured with nylon wire ties or they can be dipped in either potting compound or a wire varnish.

To reduce DCR when making your own inductor, use more than one strand of wire and parallel the ends. This has the same effect as using a single larger-gauge strand.

Resistors

Resistor quality usually has less effect on a crossover network than capacitor or inductor quality. One of the principal concerns of a resistor is that it have adequate power handling. This will be discussed in the next section.

It is easy to find capacitors and inductors with more than enough power handling for most crossover networks. However, typical carbon resistors can only handle a few watts. Often wirewound resistors are selected because high-power versions are available. But care should be taken because wirewound resistors can sometimes have significant inductance.

Finally, resistors should not be allowed to touch each other when they are mounted.

Power Handling

The most conservative approach to power handling is to use components in the crossover network that can handle the maximum continuous power that the amplifier is capable of delivering to the speaker. This should insure that no components will fail under heavy use with a continuous signal. However, if your crossover network design uses one or more resistors, this can force you to use some really huge and expensive resistors—especially if you are designing a speaker that will be connected to a high-power professional amplifier. (Many professional amplifiers can output over 1000 watts per channel!)

Many components can handle brief transient signals at high power levels as long as the average power level is much lower. Knowing the intended use of the speaker can help you determine what the average power level might be. With this in mind, many designers safely use components with much lower power handling limits.

The power handling limit of capacitors is usually rated in volts rather than watts. The following equation can estimate the peak-to-peak voltage for a desired power level:

$$V_{pp} = 2.828 \times (P \times Z)^{1/2}$$

In this equation Vpp is the peak-to-peak voltage in volts, P = the power level in watts and Z = the load impedance on the amplifier at the specified power level in ohms. For example, 100 watts into an 8 ohm load would produce 80 peak-to-peak volts.

A very rough guideline employed by some designers is to use capacitors that can handle a minimum of 50 V and resistors that can handle at least 50 W.

Calculating the maximum power handling capability of an inductor is more difficult. It depends on the length and gauge of wire and the type of insulation on it. Generally, if 18 gauge wire is used for band-pass and high-pass inductors, there should be no problem with power handling. 16 gauge wire or larger should be adequate for most low-pass inductors.

It is strongly recommend that you test your crossover network under maximum power conditions to make sure that none of its components become too hot. Crossover networks have been known to ignite a fire if they cannot handle the power.

7 Verifying & Adjusting the Design

Unless you have prior experience designing a crossover network with all of the components involved (the drivers and box) it will probably be necessary to build a prototype of the network and to test and adjust it. This is called "tweaking" and two types of tests are usually performed:

Sound Pressure Level Measurement

The most common test is to measure the sound pressure level (SPL) of the speaker with either an SPL meter or a real-time analyzer (RTA). This is usually done at several frequencies across the entire operating range of the system. Many RTAs measure the SPL at ⅓ octave intervals from 20 Hz to 20 kHz. However, SPL meters and RTAs have an important limitation. They cannot separate the direct sound of the speaker from the reflected sound that bounces off the floor, ceiling, walls and objects in the space where the measurement is made. This can introduce a variety of problems to the measurement.

A better way to measure the sound pressure level is with a time delay spectrometry (TDS) system or with a fast Fourier transform (FFT) system that uses a maximum length sequence (MLS) test signal. Depending on their settings, these more advanced measurement systems can be made to reject much of the reflected energy and measure primarily the direct sound from the speaker. This will improve the accuracy of the measurement and enable a better analysis of the crossover network.

To learn more about measurement techniques and systems, get a copy of the book *Testing Loudspeakers* by Dr. Joseph D'Appolito from Audio Amateur Press (see Appendix E).

Listening Tests

By far, the most important and the least expensive test of a speaker system is a listening test. Human hearing can still pinpoint problems much faster than instrumentation. Unfortunately, human hearing can be subjective rather than objective. Unless the listener is experienced, he or she may have difficulty discerning accuracy. For example, a common trick employed by some hi-fi speaker manufacturers is to subtly boost the high-frequency and low-frequency response of a speaker so that it "stands out" when compared to a competitor's speaker. Listeners may choose it because it sounds "brighter" and has more "bottom end". However it is less accurate.

Quick "A-B" comparisons between a sound source and a speaker that is reproducing the same sound are one way to improve the effectiveness and accuracy of listening tests. But this can be difficult to arrange.

Listening tests should still be the "final" test for a speaker system. No matter how well it measures, it still has to sound good to people.

Adjusting the Network

After measuring the speaker system you will know if the crossover network is performing as desired. Is the sound pressure level smooth through the region of each crossover frequency? Or are there dips or peaks? If there is a dip, you can redesign the crossover network with X•over Pro and create more overlap at the crossover region by adjusting the filter frequencies. If there is a peak, you can create less overlap at the crossover frequency.

Are the levels of each driver the same? If not, you may need to add or adjust an L-pad.

Example: The response has a 3 dB dip at a crossover frequency of 1000 Hz between the low-pass and high-pass filters of a two-way 4th-order crossover network. Try moving the LP filter frequency up to 1050 Hz and the HP filter frequency down to 950 Hz. (Use larger adjustments for networks with lower orders.) Plot the results in X•over Pro and observe if the response was increased 3 dB at the crossover frequency. If not, continue to adjust the filter frequencies until it does. Then, using the new component values, adjust the prototype crossover network and measure the system again to see if the problem is fixed. Continue this process until the desired results are obtained.

Tip: Experiment with X•over Pro if you are unfamiliar with how to adjust crossover network components to shape the response. You can do this by manually changing the value of a component in the component list of the main window and then plotting the response and observing how it changed. You may want to start with a simple network, like a 1st-order or 2nd-order two-way crossover network until you gain more experience.

7

X•over Pro Reference

This is the place to go when you need an explanation of one or more of the features of X•over Pro. The previous section of this manual focused on how crossover networks and their associated circuits work. This Reference section focuses on X•over Pro itself.

Chapters

1 Menus

Like many programs designed for Microsoft Windows, X•over Pro contains a menu bar across the top of the main window. Most commands can be found in the menus and sub-menus which drop down from this menu bar. In addition, the graphs have a popup menu which is displayed when you right-click (🖱) on a graph.

File Menu

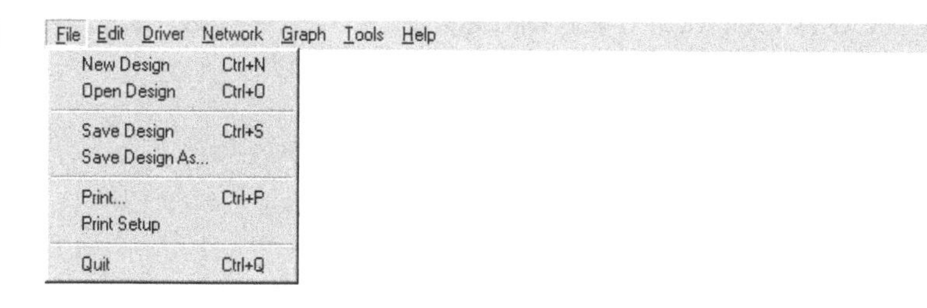

New Design

Clears an existing design and begins a new one. After the command is selected, the main window and all variables will be cleared. However, before the "New Design" command is executed, X•over Pro will check to see if there are unsaved changes and give you a chance to save them. (Keyboard shortcut: Ctrl+N.)

Open Design

Opens a crossover network design file which was previously saved to disk. You will be prompted for a file name and path. *Note: X•over Pro can also open design files from earlier versions of X•over.* (Keyboard shortcut: Ctrl+O.)

Save Design

Saves design changes made to an existing crossover network design file. You will not be prompted for a file name and path unless the design is new and has not yet been saved. (Keyboard shortcut: Ctrl+S.)

Save Design As...

Saves a new crossover network design file or a copy of an existing design with a new file name and path. The file name should end with the extension ".xo3". If it does not, X•over Pro will automatically append the extension.

1

Print...
Prints a crossover network design. A "Print an X•over Design" window will open so you can configure the printout. (Keyboard shortcut: Ctrl+P.)

Print Setup
Opens the Print Setup window so you can choose another printer and/or change the settings of the printer driver. *Important: Make sure the printer is set to print on a letter-size or A4-size page in portrait mode*. The Print Setup window can also be accessed from the "Print an X•over Design" window.

Quit
Closes the program. You will be given a chance to save any unsaved changes before the program closes. (Keyboard shortcut: Ctrl+Q.)

Edit Menu

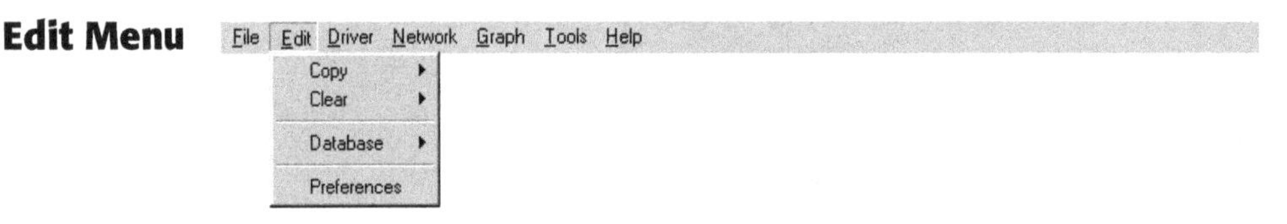

Copy > Component Values
Copies a list of the capacitors, inductors and resistors used in the crossover network, impedance equalization networks and L-pads to the Windows clipboard. Also included are the filter frequencies and filter types. The list is copied to the clipboard as text and can be "pasted" into most Windows word processing and page layout programs.

Copy > Schematic Picture
Copies a black and white picture of the schematic to the Windows clipboard. It is copied as a bitmap picture and can be "pasted" into most Windows paint programs, image editors, word processing and page layout programs.

Clear > All Circuits
Clears the crossover network and filters, all impedance equalization networks and all L-pads. The driver and box data will not be cleared. The program will ask for confirmation before executing this command to prevent accidental data loss.

Clear > Crossover / Filters Only
Clears only the crossover network and filters. The impedance equalization networks, L-pads, drivers and box data will not be cleared. The program will ask for confirmation before executing this command to prevent accidental data loss.

Clear > Impedance EQ Circuits Only
Clears only the impedance equalization networks. The crossover network, filters, L-pads, drivers and box data will not be cleared. The program will ask for confirmation before executing this command to prevent accidental data loss.

Clear > L-Pads Only
Clears only the L-pads. The crossover network and filters, impedance equalization networks, drivers and box data will not be cleared. The program will ask for confirmation before executing this command to prevent accidental data loss.

Database > Edit Driver Data
Opens the Edit Database Driver Data window so you can make changes to the drivers in the driver database. Drivers can be edited, added and deleted. (Keyboard shortcut: Ctrl+W.) *Caution: To prevent unwanted data loss it is a good idea to backup the database before editing it. To do this make a copy of file "htaudio.mdb" in the X•over Pro folder ("c:\Program Files\HT Audio\").*

Database > Edit Company Data
Opens the Edit Companies window so you can make changes to the company information in the driver database. Companies can be edited, added and deleted. *Caution: To prevent unwanted data loss it is a good idea to backup the database before editing it. To do this make a copy of file "htaudio.mdb" in the X•over Pro folder ("c:\Program Files\HT Audio\").*

Database > Compact Database
Deleting drivers from the database does not decrease its size because it would require the entire database to be resaved, greatly slowing its operation. To compact the database after one or more drivers have been deleted, select the "Database > Compact Database" command. *Note: A backup copy of the uncompacted database will be made before the database is compacted. It is named "htaudio.bak". To restore it, delete "htaudio.mdb" and rename the backup copy by changing the "bak" extension to "mdb".*

Database > Repair Database
If the computer crashes while the driver database is being edited, the driver database may become corrupted. If you ever receive an error message saying that the database is unrecognizable or is not a *Microsoft Access* database when you attempt to open it, it will probably need to be repaired. Select the "Database > Repair Database" command to repair it. *Caution: Some data may be lost when the database is repaired because corrupted portions that cannot be reconstructed will have to be discarded. Also note: A backup copy of the unrepaired database will be made before the database is repaired. It is named "htaudio.bak". To restore it, delete "htaudio.mdb" and rename the backup copy by changing the "bak" extension to "mdb".*

Preferences
Opens the Preferences window so you can edit the default settings of X•over Pro. They are saved in the X•over Pro folder in file "xopref.ini".

Driver Menu

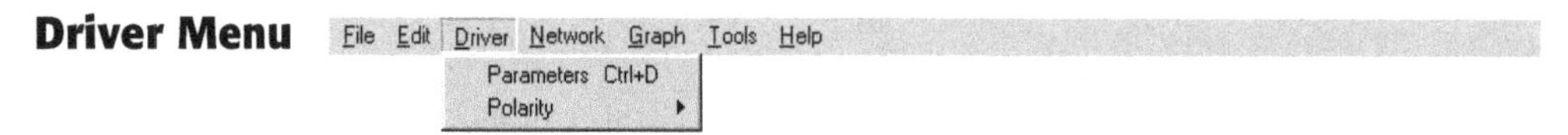

Parameters
Opens the Driver Properties window so you can enter or edit information about the woofer, midrange driver and/or tweeter. (Keyboard shortcut: Ctrl+D.)

Polarity > Invert Woofer, Midrange or Tweeter Polarity
Inverts the wiring polarity of the selected driver (woofer, midrange driver or tweeter) at its terminals. The +/– signs will be updated in the schematic. Three graphs can be affected by this option: the mini preview graph, Normalized Amplitude Response and Phase Response.

However, the mini preview and Normalized Amplitude Response graphs will not show any change unless the net response is enabled and is configured to include the effects of phase. See Chapter 11 (pages 219-220) for more information. Remember that the graphs show more than just the crossover network. They also show the box response. This includes the phase response of the box and this can also cause the net response to cancel at unexpected times when the effects of phase are included in the net response.

Network Menu

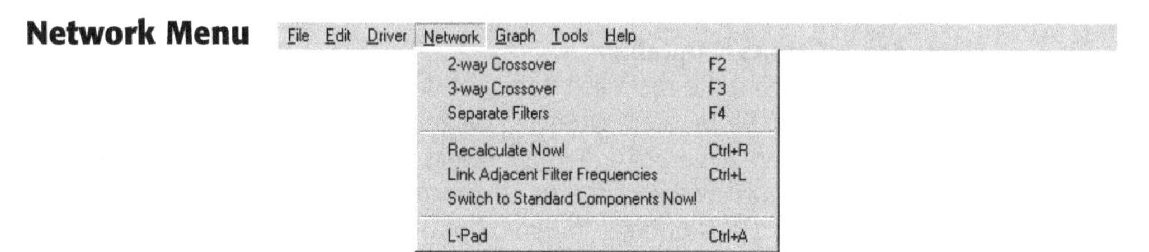

2-way Crossover
Sets the network to a two-way crossover network and configures the program accordingly. The program will assume that the tweeter is connected to the high-pass filter and the woofer is connected to the low-pass filter. The midrange driver will be ignored. (Keyboard shortcut: F2.)

3-way Crossover
Sets the network to a three-way crossover network and configures the program accordingly. The program will assume that the tweeter is connected to the high-pass filter, the midrange driver is connected to the band-pass filter and the woofer is connected to the low-pass filter. (Keyboard shortcut: F3.)

Separate Filters

Sets the network to "Separate Filters" and configures the program accordingly. A window will open to prompt you to assign a driver to each of the filters that you plan to use and, since each filter is assumed to be separate, the net response option in the graphs will be disabled. *Note: It is possible to assign the same driver to more than filter. However, X•over Pro treats each filter as if it has its own driver. If, for example, the woofer is selected for both the low-pass and band-pass filters, the program will assume that each filter has its own copy of the woofer. This is readily apparent in the graphs because they plot each filter separately with no net response option.* (Keyboard shortcut: F4.)

Recalculate Now!

Causes the program to immediately recalculate the component values of all capacitors, inductors and resistors in the crossover network or filters. The component list in the main window will be updated. This command will not affect the impedance equalization networks or L-pads. (Keyboard shortcut: Ctrl+R.)

Use the "Recalculate Now!" command after a driver has been changed or configured differently. Also use it whenever an adjacent filter has been changed and you want to insure that all other filters are also refreshed.

Link Adjacent Filter Frequencies

Normally, adjacent filter frequencies in a crossover network should be the same. This can be handled automatically with the "Link Adjacent Filter Frequencies" command. When this feature is enabled, changes made to one filter frequency will be automatically mirrored in the adjacent filter frequency. This command is not available when the network is set to "separate filters." (Keyboard shortcut: Ctrl+L.)

Switch to Standard Components Now!

X•over Pro always tries to calculate precise values for components. Unfortunately a component may not be available with a desired value. There are two ways to deal with this. One way is to combine two or more similar components to achieve the desired value. For example, two 4.7 μF capacitors can be paralleled to achieve a net value of 2.35 μF. A "Parallel-Series Value Calculator" is included in the Tools menu to aid with such calculations.

The second way of dealing with this problem is to substitute the closest standard-value component and see if the result is satisfactory. The "Switch to Standard Components Now!" command is provided to do just that. It will immediately substitute the closest standard value for all capacitors, inductors and resistors in the crossover network or filters, impedance equalization networks and L-pads.

If the standard values are unsatisfactory, how do you restore the precise values again? Execute the "Recalculate Now!" command to recalculate the crossover network components. Return to the Driver Properties window and use the "Calculate EQ Network" button on the

1

"Impedance EQ" tab to restore each impedance equalization network. And return to the L-pad window to restore each L-pad.

Note: The standard values are gleaned by Harris Tech from electronics component manufacturers and catalogs.

L-Pad
Opens the L-Pad Attenuator window so you can add, remove or edit an L-pad. (Keyboard shortcut: Ctrl+A.)

Graph Menu

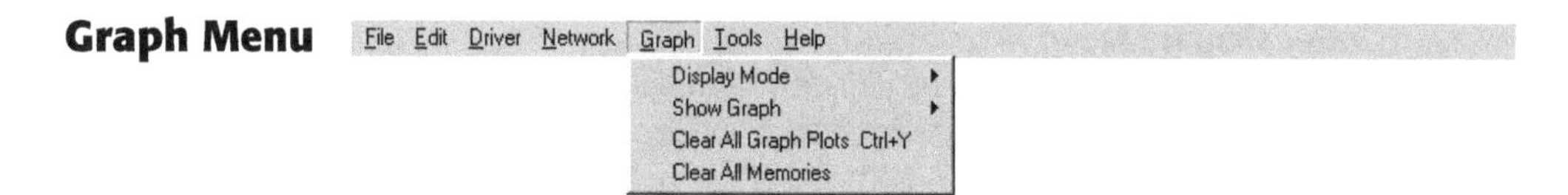

Display Mode
Two display mode commands are provided: "Single Window" and "Individual Windows". The "single window" mode combines all graphs into a single window where they can be individually accessed with a tab button. This mode is most useful for lower screen resolutions such as VGA and SVGA because it economizes space. The "single window" mode also uses less memory and system resources. In the "individual windows" mode each graph opens in its own window so multiple graphs can be viewed at the same time. The "individual windows" mode also provides a size option to allow each graph window to be enlarged to a fixed "large" size.

Show Graph > Amplitude–Normalized
Displays the Normalized Amplitude Response graph. This is often referred to as the "frequency response". It shows the relative level differences between each filter in a crossover network and it includes the effects of the drivers and box. (Keyboard shortcut: Ctrl+F1.)

When a two-way or three-way crossover network is selected, the Normalized Amplitude Response graph includes a net response option. The net response can be calculated one of two different ways depending on the setting of the "ø" (phase) control in the "Graph" tab of the Preferences window. When "ø" is turned on, the net response is calculated by summing both the magnitude and phase of the volume velocity of each filter. *(Important: This assumes that all drivers are carefully mounted with their acoustical centers located close together and aligned so that the direct sound arrives from each driver to the listener's ear at the same time. The driver mounting you choose can drastically alter the actual net acoustical response that you receive. This is one reason why the second net response calculation method is provided.)*

When the "ø" control is turned off, the net response will be calculated by summing only the

1

magnitude of each filter. Phase cancellations will be ignored with this method.

Show Graph > System Impedance
Displays the System Impedance graph. It shows the impedance that the amplifier will see after considering the effects of the crossover network or filters, impedance equalization networks, L-pads, drivers and box. (Keyboard shortcut: Ctrl+F2.)

When a two-way or three-way crossover network is selected, the System Impedance graph includes a net impedance option. It is calculated by combining the impedance of each parallel filter.

Show Graph > Phase
Displays the Phase Response graph. It shows how far the output signal from each driver will lag behind the input signal. The signal delay is displayed as a phase shift. (Keyboard shortcut: Ctrl+F3.) *Important: If box information has been entered, the phase response includes the effects of the box.*

Show Graph > Group Delay
Displays the Group Delay graph. It shows how far the output signal from the speaker will lag behind the input signal. The signal delay is displayed in milliseconds. (Keyboard shortcut: Ctrl+F4.) *Important: If box information has been entered, the group delay includes the effects of the box.*

Clear All Graph Plots
Clears the plot lines from all graphs. (Keyboard shortcut: Ctrl+Y.)

Clear All Memories
Clears all seven of the graph memories.

Tools Menu

File Edit Driver Network Graph Tools Help
Component Resistance Estimator Ctrl+E
Parallel-Series Value Calculator
Color Value Decoder
Notch Filter Designer
Start BassBox Pro

Component Resistance Estimator
Opens the Filter Component Resistance Estimator window to estimate the ESR (equivalent series resistance) of capacitors and/or DCR (DC resistance) of inductors. (Keyboard shortcut: Ctrl+E.) *Note: X•over Pro needs the ESR and DCR for the capacitors and inductors of the crossover network filters only.*

Parallel-Series Value Calculator
Opens the Parallel-Series Value Calculator window to calculate the resultant value of multiple resistors and/or capacitors.

Color Value Decoder
Opens the Color Value Decoder window to decode the value of a resistor, capacitor or inductor from it color bands.

Notch Filter Designer
Opens the Notch Filter Designer window to design a separate series or parallel notch filter. *Note: Filters designed with the Notch Filter Designer are not included in the schematics, graphs or printouts.*

Start BassBox Pro
Launches the BassBox Pro program if you have purchased a BassBox Pro license and the program is installed on your computer. BassBox Pro is a speaker box design program from Harris Tech (see pages 209-210).

Help Menu

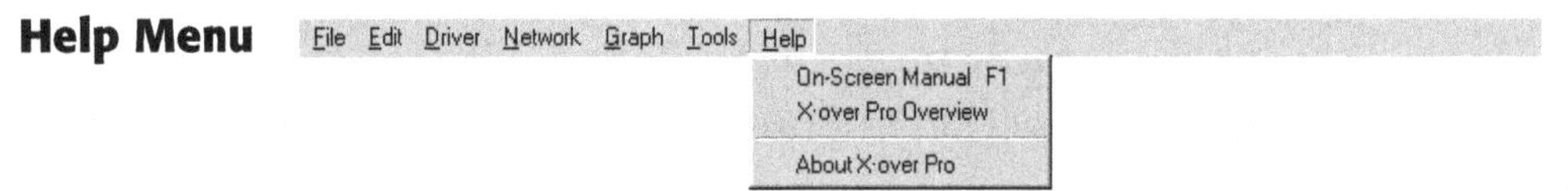

On-Screen Manual
Opens the on-screen manual to its Contents page. From it you can select from several topics. (Keyboard shortcut: F1.)

X•over Pro Overview
Opens the on-screen manual to its "An Overview of X•over Pro" topic which introduces first-time users to the program.

About X•over Pro
Opens the X•over Pro title window and displays the program's serial number, user registration information and copyright notice.

1

Graph Popup Menu

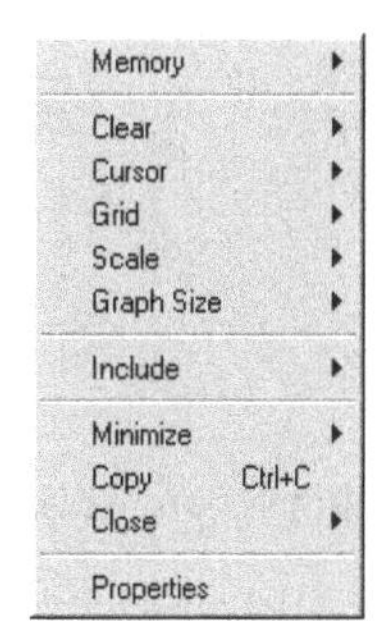

Memory > Store
Stores a "snapshot" of the most recently plotted design into one of seven graph memories so it can be replotted again later. The design parameters are stored into memory rather than the plot line data points. This enables all open graphs to be replotted when a memory is recalled—even if the graphs were closed when the design was originally stored into the graph memory. (Keyboard shortcuts: [⇧Shift]+[F1] to [F7].)

Memory > Recall
Replots one or all designs stored in the graph memories. (Keyboard shortcuts: [⇧Shift]+[Ctrl]+[F1] to [F8].)

Memory > Clear All
Clears all seven of the graph memories

Clear
Clears all plot lines from the selected graph or all open graphs. (Keyboard shortcuts: [Ctrl]+[X] to clear the selected graph only and [Ctrl]+[Y] to clear all graphs.)

Cursor > Show in Selected Graph
Activates the cursor in the selected graph. (Keyboard shortcut: [Ctrl]+[U].)

Cursor > Show in All Graphs
Activates the cursors in all graphs.

Cursor > Link All Cursors
Links the cursors in all graphs ("individual windows" mode only) so they will move as one.

Cursor > Unlink All Cursors
Unlinks the cursors in all graphs ("individual windows" mode only) so they can be controlled separately.

1

Cursor > Hide in Selected Graph
Deactivates the cursor in the selected graph. (Keyboard shortcut: Ctrl+H.)

Cursor > Hide in All Graphs
Deactivates the cursor in all graphs.

Grid
Displays or removes the grid in the selected graph or all graphs.

Scale > Vertical: Normal
Displays the standard vertical scale in the selected graph.

Scale > Vertical: Expanded
Displays an expanded vertical scale in the selected graph.

Graph Size
Two sizes are provided ("individual windows" mode only): normal and large. The large size does not increase the range of either the vertical or horizontal scales but it does add more data points, improving the detail of the plot lines. This is because the plotting functions in X•over Pro calculate a graph data point for each pixel column in the graph.

Include > Driver Acoustic Response
Includes the driver acoustic response—if it has been entered. Only the mini preview and Normalized Amplitude Response graphs are affected by this option. (Keyboard shortcut: Ctrl+I.)

Include > Vent Resonance Peaks
Includes "pipe" resonance peaks. These peaks are a natural result of the vent in vented and bandpass boxes. All graphs are affected by this option. The required vent data for this feature can only be imported from a BassBox Pro or BassBox Lite speaker design file—it cannot be created in X•over Pro.

Include > Room/Car Acoustic Response
Includes the acoustic response of the listening space—if it has been entered. Only the mini preview and Normalized Amplitude Response graphs are affected by this option. The required acoustical data for this feature can only be imported from a BassBox Pro design file—it cannot be created within X•over Pro.

Include > On-Axis Piston Band Response
Includes the estimated on-axis piston band response—if the diaphragm diameter (Dia) or diaphragm area (Sd) of the driver has been entered. Only the mini preview and Normalized

Amplitude Response graphs are affected by this option. The "piston band" is the frequency band where the driver maintains a constant load versus frequency. This begins at the frequency whose wavelength is equal to the circumference of the diaphragm (piston) and it extends upward in frequency. In the piston band, the on-axis response of most drivers will begin to increase with frequency at a rate of 6 dB/octave because the beamwidth or coverage angle of the driver will gradually narrow as the frequency increases.

Include > Diffraction Response Shelf
Includes the generalized effects of diffraction for a limited selection of box shapes (cube, square prism and optimum square prism). Diffraction is the bending of sound waves as they pass near an edge or corner of a solid object. Only the mini preview and Normalized Amplitude Response graphs are affected by this option. The required box data for this feature can only be imported from a BassBox Pro or BassBox Lite design file—it cannot be created within X•over Pro.

Include > Net (Composite) Response
The "net" response shows the overall response of the crossover network and the system. It is calculated one of two different ways, depending on the setting of the "ø" (phase) checkbox in the Preferences window (see pages 219-220). When "ø" is turned on, it is calculated by summing both the magnitude and the phase of the volume velocities of each filter. When "ø" is turned off, it is calculated by summing only the magnitudes. (Keyboard shortcut: Ctrl+Z.)

Remember that the graphs show more than just the crossover network. They also show the box response. This includes the phase response of the box and this can also cause the net response to cancel at unexpected times when "ø" is turned on.

Important: X•over Pro must make a few assumptions when "ø" is turned on. One assumption is that all drivers are carefully aligned so that the sound from each one arrives at the listener's ear at the same time (this is a goal of many, if not most, designs). However, many systems have misaligned drivers so the final amplitude and phase response can vary significantly from the model. This is one reason why it is important to prototype, test and adjust a design before finalizing it.

The graphs affected by this option are: the mini preview graph, the Normalized Amplitude Response graph and the System Impedance graph. The Phase Response and Group Delay graphs do not offer a net plot line option.

Although the graph popup menu and Graph Properties window only allow the net plot line to be turned on and off, it is also possible to turn on and off the individual filter plot lines. To do this use the "Graph" tab of the Preferences window.

1

Minimize
Hides either the selected graph or all graphs and places a graph icon on the Windows desktop. To restore a graph window, click on its icon.

Copy
Creates a copy of the selected graph in the Windows clipboard so that it can be pasted into another application (like a word processor or page layout program). This command actually "prints" to the clipboard, creating a printable version of the graph with a white background. The color of the plot lines is preserved, except for white which is switched to black. Be aware, however, that some colors like yellow may be very light. (Keyboard shortcut: Ctrl+C.)

Close
Closes either the selected graph or all open graphs.

Properties
Opens the Graph Properties window, displaying the graph settings.

2 Driver Properties

The first step in designing a crossover network is to enter information about the drivers. This is accomplished with the Driver Properties window shown below. Notice that it uses a tab layout to organize the driver information.

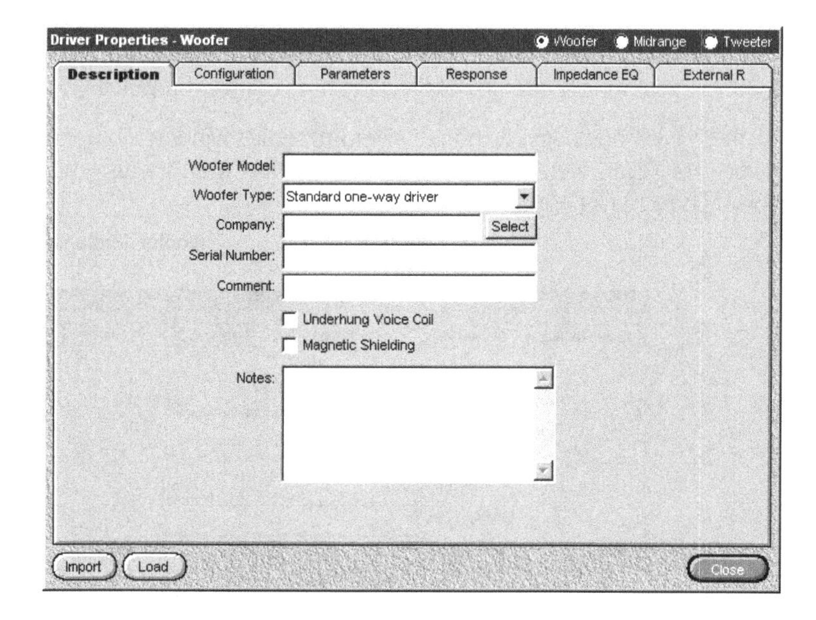

Note: The tabs of the Driver Properties window show the information for only one driver at a time (woofer, midrange driver or tweeter).

The importance of entering accurate information cannot be overstated. The accuracy and capabilities of X•over Pro are directly related to the quality and quantity of the driver information. Three ways are provided to enter driver information:

1 **Driver Database** X•over Pro includes a huge driver database with information for thousands of drivers. This is the first place to look for driver information. The driver database is opened with the "Load" button at the bottom of the Driver Properties window. It can be easily searched by manufacturer, model, driver parameters or box parameters. Once a driver is located, its information is "loaded" into X•over Pro by clicking on the "Load" button of the Driver Locator window.

2 **Import Parameters** X•over Pro also provides limited import capability for driver Thiele-Small parameters. These parameters can be imported from a "txt" (T-S) file created with the CLIO measurement system, a "txt" (T-S) file created with either a DATS (Dayton Audio Test System) or WT3 (Dayton Audio Woofer Tester 3), an "ats" file created with the LAUD version 3.12 (or later) measurement system, or a "log" file cre-

ated with the "'Export to BassBox" command of a WT2 (Smith & Larson Audio Woofer Testrer 2) measurement system.. They can also be imported from a BassBox 6 Pro/ Lite "bb6" speaker design file. The import feature is accessed with the "Import" button at the bottom of the Driver Properties window.

3 **Manual Entry** Driver information can be entered manually by the user. This is a convenient way to enter data direct from a manufacturer's data sheet or website. It also allows the user to enter measured Thiele-Small parameters. Information can be entered manually into any of the tabs of the Driver Properties window.

2

All three methods begin with the Driver Properties window. To open it select the "Parameters" command from the Driver menu or use the keyboard shortcut Ctrl+D.

Woofer, Midrange & Tweeter selection buttons

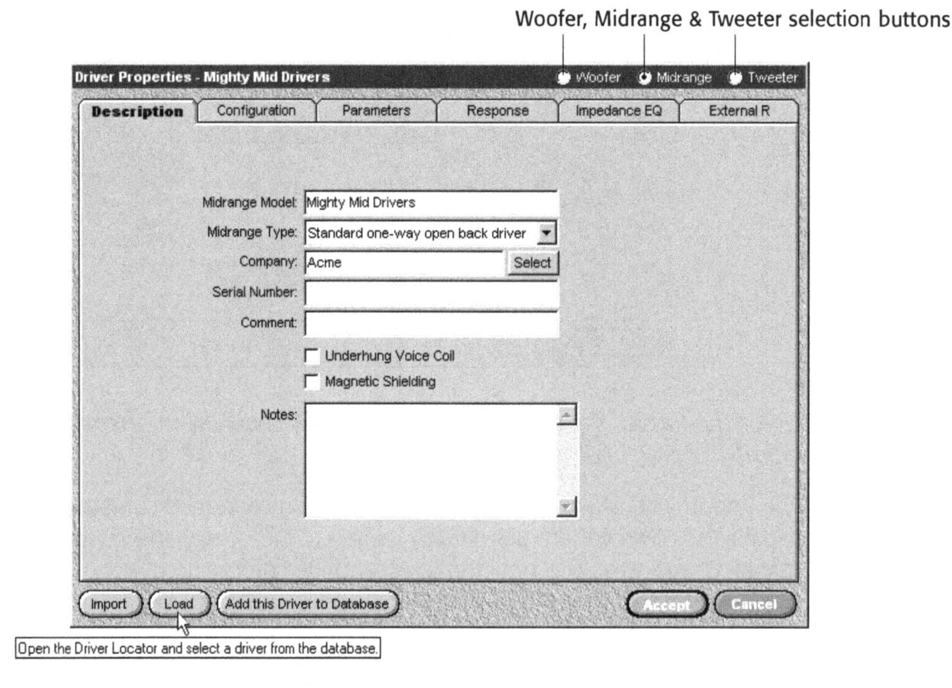

Next, let's discuss the overall use of the Driver Properties window and its general controls.

Woofer, Midrange, Tweeter Since X•over Pro can design three-way crossover networks it allows you to enter up to three different drivers: a woofer, midrange driver and tweeter. However, the Driver Properties window can display only one driver at a time. This is why the "Woofer", "Midrange" and "Tweeter" buttons are provided in the title bar in the upper right corner of the window. Use these buttons to select a driver so its parameters can be entered, edited or displayed in the window. Notice also that the name of the selected driver will be appended to the name of the window on the left end of the title bar.

Note: Midrange drivers should only be entered for three-way crossover networks or a separate filter. Midrange driver information is ignored for two-way crossover networks.

When a driver is selected, X•over Pro makes certain assumptions about it. These assumptions are summarized next:

Open Back versus Sealed Back

All woofers are assumed to have an open back while all tweeters are assumed to have a sealed back. However, midrange drivers can fall into either category and this is selected with the "Midrange Type" drop-down list on the "Description" tab of the Driver Properties window. See the Chapter 2 of the *Crossover Network Designer's Guide* earlier in this manual for more information about open back and sealed back drivers.

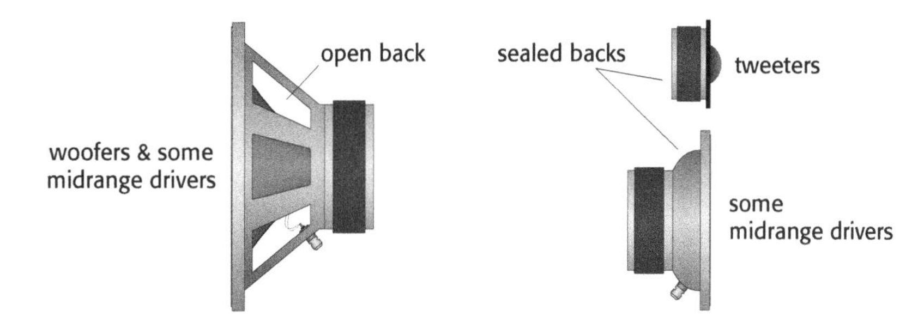

Use the "Midrange Type" drop-down list to identify the midrange driver as an "open back" or "sealed back" driver.

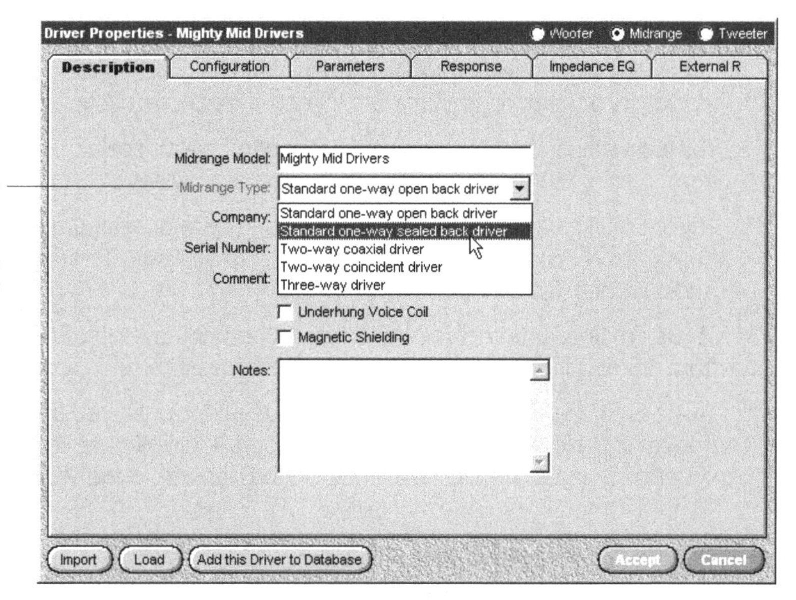

2

Filter Connections (Driver Assignments)
The woofer is always assumed to be connected to the low-pass filter in a two-way or three-way crossover network. The tweeter is always assumed to be connected to the high-pass filter in a two-way or three-way crossover network. The midrange driver is always assumed to be connected to the band-pass filter in a three-way crossover network. However, X•over Pro allows you to select any driver for any filter if "Separate Filters" is selected in the Network menu of the main window. *Note: When the "Separate Filters" setting is used, the graphs assume that each filter has its own separate driver even if the same driver is selected for more than one filter. Also, the "net" or composite plot line feature will be disabled in the graphs.*

Several buttons (shown below) are provided at the bottom of the Driver Properties window. Some of them may or may not be available depending on the status of the window or the driver being displayed in it. Each button is described next.

Import This button provides a way to import information from five different sources:

1 Information about an open back driver, its box and the acoustic listening environment can be imported from a BassBox 6 Pro or BassBox 6 Lite "bb6" speaker design file. (BassBox 6 Pro and Lite are speaker box design programs from Harris Tech.) Depending on the amount of information available in the bb6 file, the driver information can include its description, parameters, dimensions and acoustic response.

2 Thiele-Small parameters for any driver (open back or sealed back) can be imported from a "txt" (T-S) file from a CLIO measurement system.

3 Thiele-Small parameters for any driver (open back or sealed back) can be imported from a "txt" (T-S) file from either a DATS (Dayton Audio Test System) or WT3 (Dayton Audio Woofer Tester 3).

4 Thiele-Small parameters for any driver (open back or sealed back) can be imported from an "ats" file from a LAUD 3.12 or later measurement system.

5 Thiele-Small parameters for any driver (open back or sealed back) can be imported from a "log" file from a WT2 (Smith & Larson Audio Woofer Tester 2) measurement system. To create this file, select "Export to BassBox" in the Woofer Tester 2's "WooFile" menu.

When the "Import" button is clicked an Import Driver T-S Parameters window will appear. Use the "Files of type" list to select the file type you want to import. *Note: The location of the "Files of type" list varies with the version of Windows. On Windows 8, 7 and Vista, it is not labeled and is located to the right of the "File name" input box. In older version of Windows, it is labeled and located at the bottom, below the "File name" input box.*

The BassBox 6 files end with the extension "bb6", the CLIO, DATS and WT3 files end with the extension "txt", the LAUD files end with the extension "ats" and the WT2 files end in the extension "log". *Note: Depending on how the file settings of Microsoft Windows are configured, you may or may not see these file name extensions at the end of each file name.*

Load Opens the Driver Locator window so the driver database can be searched for a driver. This button is hidden while the Driver Locator window is open. Once a driver is found it can be "loaded" into a design (using the "Load" button on the Driver Locator window). When a driver is loaded all of its information will be copied to the Driver Properties window.

Add this Driver to Database This button is available whenever sufficient driver information has been entered into the Driver Properties window. It is used to add a driver to the database and it is provided as a convenience for those occasions when new drivers are entered. However, there is a better way to add to or edit the driver database (see Chapter 9).

Close The "Close" button is available whenever no changes have been made. Use it to close the Driver Properties window.

Accept, Cancel The "Accept" and "Cancel" buttons replace the "Close" button whenever changes are made to any of the drivers while the Driver Properties window is open. Use the "Accept" button to accept changes and close the Driver Properties window. Use the "Cancel" button to restore the previous driver properties and close the Driver Properties window.

Next, let's discuss each of the three methods of entering driver information that were introduced on pages 77-78. Afterward, the remainder of this chapter will explain the controls and input boxes of each tab of the Driver Properties window.

2

Loading Driver Data from the Driver Database

The Driver Locator window (shown below) is used to locate driver information in the driver database and load it into a crossover network or filter design. (See Chapter 9 if you want to learn how to edit information in the driver database.)

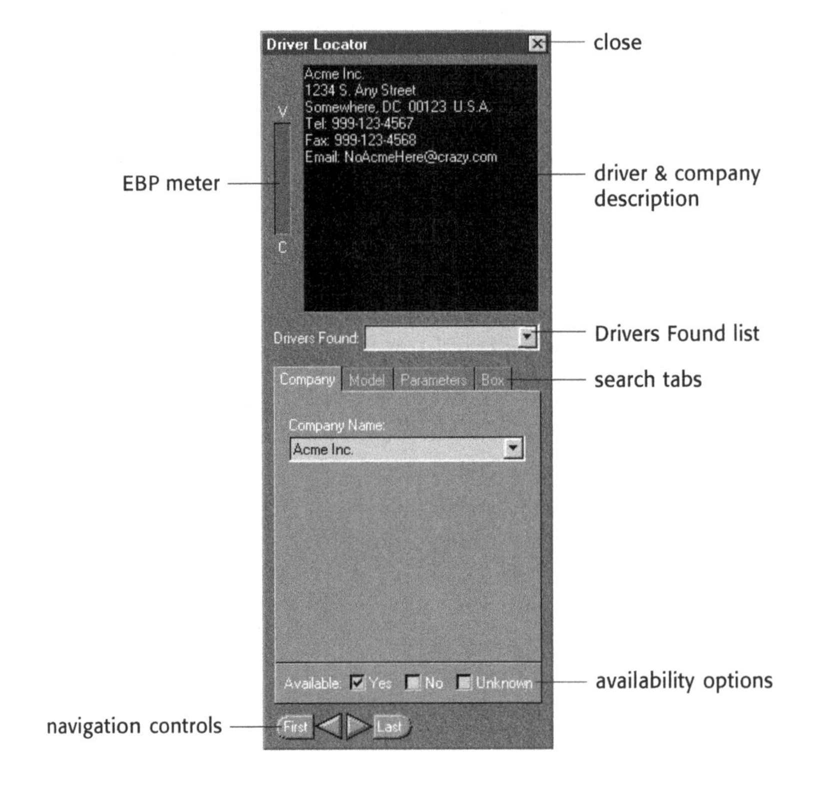

X•over Pro includes both open back and sealed back drivers (pages 41-42) in its driver database. However, it will display only one driver type at a time depending on the driver type selection. For example, if the "Tweeter" button is selected in the title bar of the Driver Properties window, the Driver Locator will show <u>only</u> sealed back drivers.

There are two places where you can select the driver type that is displayed by the Driver Locator: 1) with the "Woofer", "Midrange" and "Tweeter" buttons in the title bar of the Driver Properties window (pages 78-79) or 2) with the "Open Back" and "Sealed Back" buttons in the title bar of the Edit Database Driver Data window (pages 190-191). In both cases remember that a midrange driver can have either an open or sealed back, depending on the "Midrange Type" setting on the "Description" tab of either of these windows.

Next, let's discuss each of the objects and controls of the Driver Locator:

EBP meter (open back drivers only) The EBP meter shows the suitability of an open back driver for either a closed box or a vented box. If an open back driver is more suited for a vented box, the indicator will appear high on the meter near the "V" (vented) label at the top. Conversely, if an open back driver is more suited for a closed box, the indicator will appear low on the meter near the "C" (closed) label at the bottom. The EBP meter will disappear when the Driver Locator is switched to sealed back drivers.

EBP is an acronym for efficiency bandwidth product and it is calculated by dividing Qes (the driver's electrical Q) into Fs (the driver's free-air resonance). If Qes is unknown, X•over Pro will estimate EBP with Fs / Qts. A driver is considered better suited for a closed box when its EBP is 50 or less and better suited for a vented box when its EBP is 100 or more. Pausing the mouse pointer over the EBP indicator will cause its value to appear as shown at right.

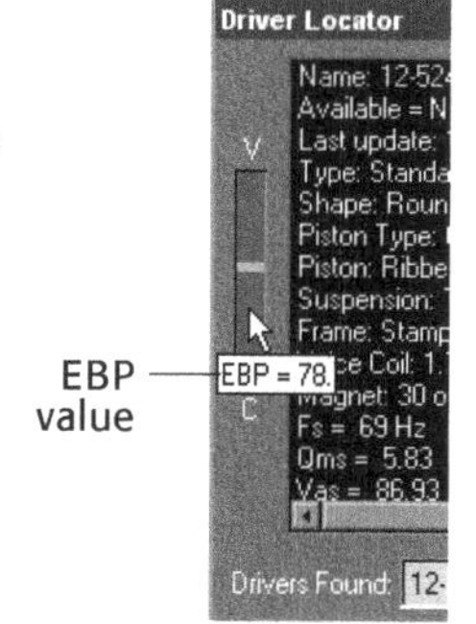

The EBP meter is only a general guide—it is not a rule. There are many exceptions where, for example, a driver with a low EBP will still work well in a vented box and visa versa.

Navigation controls The navigation controls at the bottom of the Driver Locator are rarely used during a search because they step through the database one record at a time. The "First" button selects the first driver record in the database. The "<" button selects the previous driver record in the database. The ">" button selects the next driver record in the database. The "Last " button selects the last driver record in the database.

These controls are useful when the Edit Database Driver Data window is open and you need to step through the driver records. **Caution:** Using one of the navigation controls will cause the results of a search (which are located in the "Drivers Found" list) to be cleared.

The normal way to choose a driver after a search is to select it from the "Drivers Found" list.

Close The "Close" button serves two purposes. The first is obvious—it closes the Driver Locator window. The second purpose is not obvious—it requires the use of the [⇧Shift] key and it toggles the size of the Driver Locator. We'll explain.

The Driver Locator window is available in two sizes as shown on the next page. You can switch between the small and large sizes by holding down the [⇧Shift] key and clicking the "Close" button. Select the size that looks best with your computer display. *Note: Most of the illustrations in this manual use the large size.*

Driver & company description Information about the selected company and driver are listed in this text box. The driver information, including its parameters, are always listed first. If no driver has been selected, only the company will be listed.

small size

large size

Drivers Found list After a search has been run, the resulting drivers will be listed in the "Drivers Found" list. Use this list to select a driver to be viewed in the "Driver & company description" text box above. After a driver has been selected, a "Load" button (shown below) will appear in the lower right corner of the Driver Locator window.

Clicking on the "Load" button will cause the selected driver to be loaded into the design and the Driver Locator window will close.

Search tabs There are four ways to search the driver database and each method is represented with a tab. You can search by company, model and driver or box parameters. Each of these search methods will be discussed in detail on the following pages. *Note: The fourth search method (search by box parameters) is only available for open back drivers.*

Availability options The driver database includes both old and new drivers. Having the parameters of old drivers is useful in case you ever need to design a new box for an old driver. However, most of the time you will probably want to use a driver that is currently available. For this reason, each driver in the database has a setting which identifies its availability. You can use the "Availability" search options to control which ones will be included in a search.

For example, check only the "Yes" checkbox if you only want to include drivers that are currently available (available = yes). Check all three checkboxes if you want to include all drivers in the search.

Note: The reason there is an "Unknown" setting is because there are some drivers whose availability is unknown. Contact Harris Tech at support@ht–audio.com if you have more information about a driver whose availability is unknown.

Search by Company

This is one of the easiest search methods. With the "Company" search tab selected, choose a company name from the "Company Name" drop-down list. A message window will open stating how many drivers were found for the selected company as shown below.

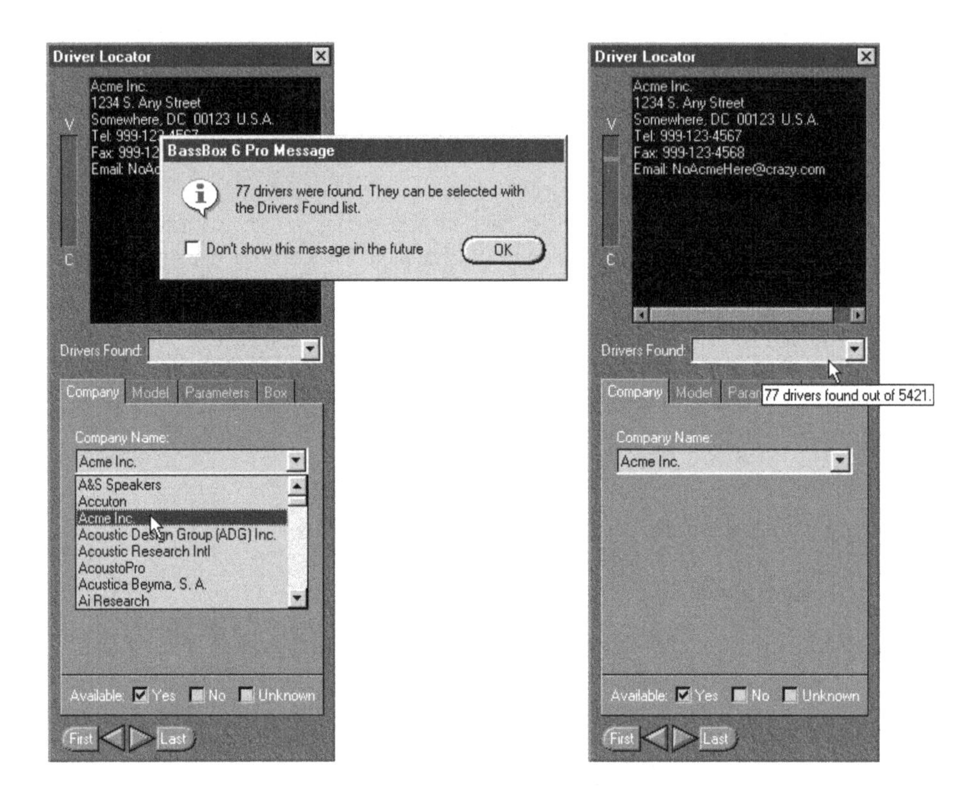

After a search, moving the pointer over the "Drivers Found" list will cause the number of drivers found and the total number of relevant drivers in the database to be listed. *Note: In X•over Pro the total number of drivers includes the total of both the open back and sealed back drivers. This makes the total larger than the figure reported by BassBox Pro..*

Immediately after a search, no drivers have been selected yet so only the company information is listed in the description box at the top of the window. Use the "Drivers Found" drop-down list to select a driver. This is shown on the next page.

Caution: Do <u>not</u> use the manual navigation controls (First, <, >, Last) at the bottom of the window after a search. Doing so will clear the "Drivers Found" list and deselect the company. Use them when you want to step through the database one record at a time.

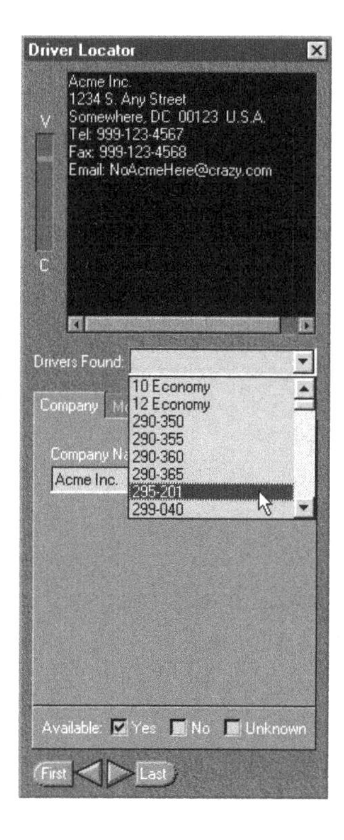

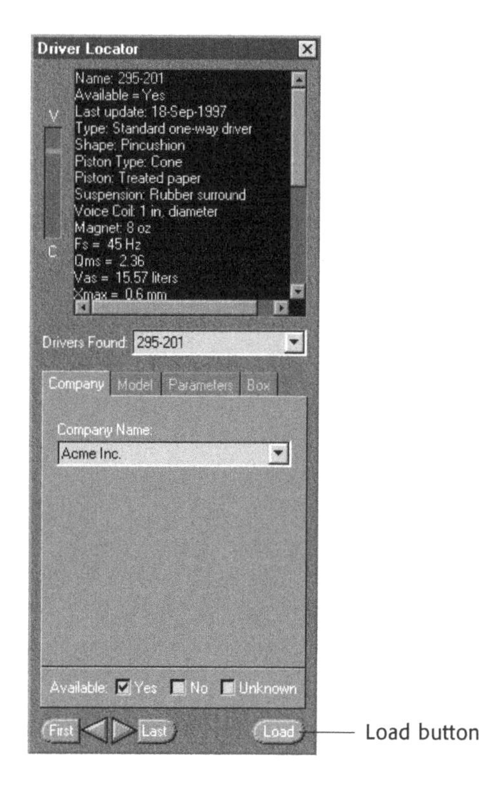

All of the drivers that were found during the search will be listed in the "Drivers Found" list. Selecting one of them causes its description, parameters and company information to be displayed in the description box. However, it does not load the driver into the crossover network design or the Driver Properties window.

Selecting a driver from the "Drivers Found" list also causes the "Load" button to appear in the bottom right corner of the window. Clicking the "Load" button will load the driver into the Driver Properties window and close the Driver Locator window.

2

Search by Model Name

This is another easy-to-use search method. With the "Model" search tab selected, enter a model name into the "Model Name" input box. Then click the "Search" button at the bottom of the window.

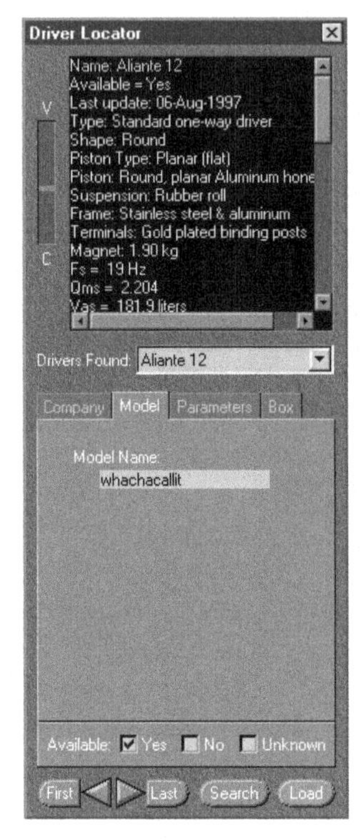

The model name is not case sensitive. The search engine will find all drivers whose model name begins with the letters and/or numbers you enter. For example, entering a single letter like "a" will produce a list of every driver beginning with "a" or "A" such as:

A11EC80-02F	A13WG-01-04 - LF
A13WH-01-04 - LF	A17WG-01-04 - LF
A25FU14-57F-Q	A25F020-53F

Wildcard searches are not supported. Any characters entered into the "Model Name" box like hyphens, asterisks, commas and periods are treated literally. However, spaces are ignored.

Search by Driver Parameters

This search method is very flexible and offers many ways to narrow the search. However, it requires more decision making on the user's part. With the "Parameters" search tab selected, begin by selecting the driver parameters you want to use in the search and then configure their units. Up to five driver parameters can be specified. Changing parameters or their units is accomplished by clicking on the parameter labels and their respective unit labels. Any of the twenty-one driver parameters used in X•over Pro can be selected (Fs, Qms, Vas, Cms, Mms, Rms, Xmax, Xmech, Dia, Sd, Vd, Qes, Re, Le, Z, BL, Pe, Qts, ηo, 1-W SPL, 2.8-V SPL). In the illustration below, Fs, Dia, Cms, Pe and Qts are selected. For example, to change Cms to Vas, simply click the Cms label. Each time a parameter label is clicked, it will change to the next available parameter.

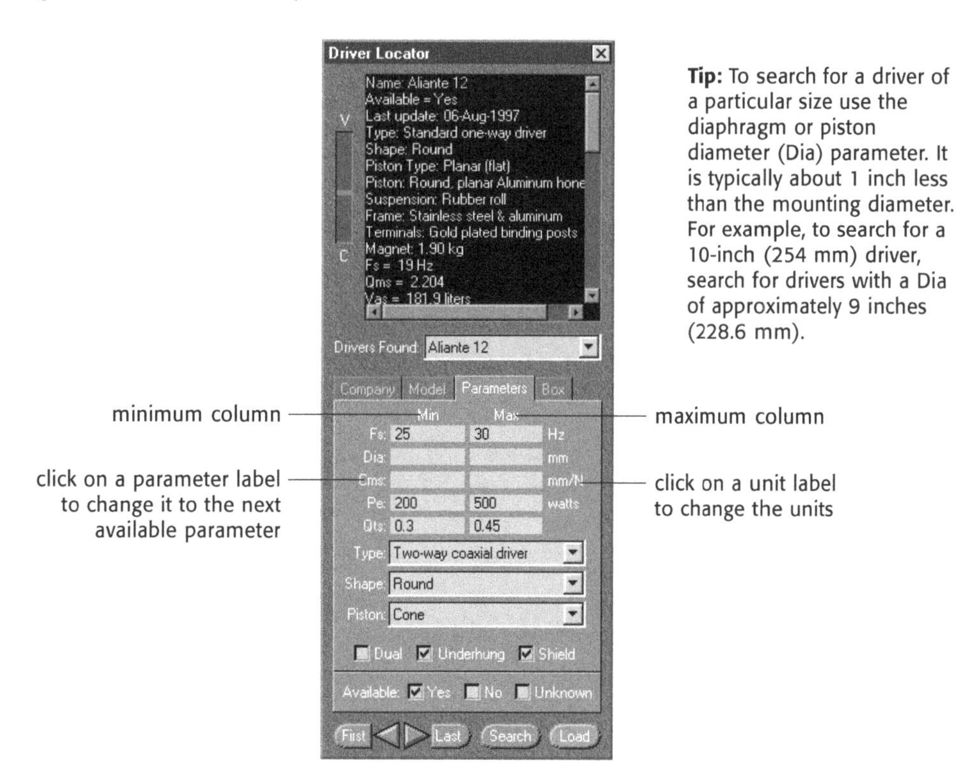

Tip: To search for a driver of a particular size use the diaphragm or piston diameter (Dia) parameter. It is typically about 1 inch less than the mounting diameter. For example, to search for a 10-inch (254 mm) driver, search for drivers with a Dia of approximately 9 inches (228.6 mm).

Next, enter a minimum ("Min") and maximum ("Max") value for each parameter. Selected parameters will be ignored during a search if a minimum and maximum value have not been entered for them. In the example above, Dia and Cms will be ignored. Remember to set the units before entering the minimum and maximum values.

2

Select the desired driver type, shape and piston type from their respective drop-down lists. Finally, include "Dual" if you want to search for a dual-voice coil driver (open back only). Select "Underhung" if you want to search for a driver (open back only) with a voice coil that is shorter than the gap of the magnet. (Underhung voice coils can behave with more linearity than a typical overhung voice coil but they are less common because they tend to be more expensive–especially if a long excursion is also required.) Select "Shield" if you want to search for a driver (either open back or sealed back) with magnetic shielding so it can be used near video monitors, computer monitors or televisions.

Click the "Search" button at the bottom of the window to begin the search.

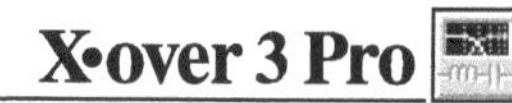

Search by Box Parameters

This search method is presently available only for open back drivers and only for closed and vented box types. With the "Box" search tab selected, choose a box type from the "Box" list. If necessary, change the box volume (Vb) units by clicking on the unit label ("cu.ft" in this example). Then enter the minimum ("Min") and maximum ("Max") box internal volume for parameter Vb. Next, enter a minimum and maximum –3 dB cutoff frequency for parameter F3. Finally, click on the "Search" button to begin the search.

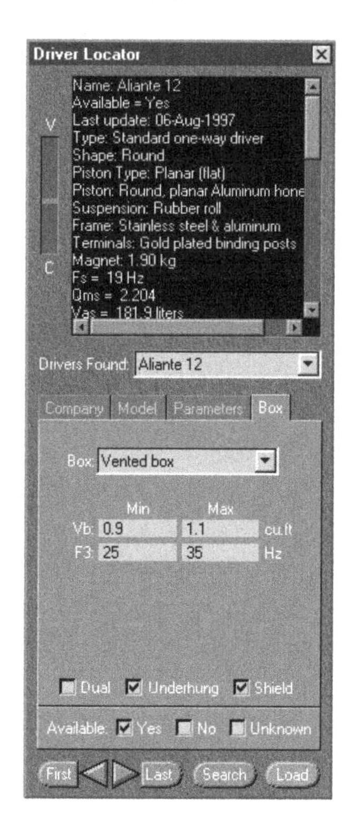

The search engine will attempt to find all open back drivers that produce a smooth amplitude response with the specified internal box volume and achieve a cutoff frequency in the specified range. If a closed box type was selected, drivers will be rejected if EBP is greater than 75 or Qts is less than 0.3. If a vented box type was selected, drivers will be rejected if EBP is less than 75 or Qts is greater than 0.5.

2

Importing Driver Data

Driver information can be imported into X•over Pro using the "Import" button at the bottom of the Driver Properties window (page 80). Import filters are provided for BassBox 6 Pro / Lite speaker design files, CLIO measurement files, DATS or WT3 measurement files, LAUD measurement files, and WT2 measurement files.

The BassBox import filter will import information about an open back driver from a BassBox 6 Pro or BassBox 6 Lite speaker design file. BassBox Pro and Lite are both speaker box design programs from Harris Tech. Since they do not use sealed back drivers, you can only import woofers and open back midrange drivers from them. For this reason, the BassBox 6 import filter is not available when the selected driver in X•over Pro is a tweeter or a sealed back midrange driver. BassBox 6 design files end with the file name extension "bb6".

Importing a driver from BassBox 6 Pro or Lite has a big advantage because it can import much more than just Thiele-Small parameters. If available, it can also import box data, cable resistance and the acoustical response curves of the driver and listening environment. The box information is more complete because it includes vent dimensions and other information that is not accessible from within X•over Pro. The acoustical response of the listening environment is also not available from within X•over Pro. However, this additional information can still be utilized in the graphs to display a more complete speaker system response.

The CLIO import filter will import driver parameters from a "txt" (T-S) file from the CLIO measurement system by Audiomatica S.R.L. The CLIO import filter is available for both open back and sealed back drivers.

The DATS / WT3 import filter will import driver parameters from a "txt" (T-S) file from the DATS (Dayton Audio Test System) or WT3 (Dayton Audio Woofer Tester 3). The DATS / WT3 import filter is available for both open back and sealed back drivers.

The LAUD import filter will import driver parameters from an "ats" file from the LAUD measurement system by Liberty Instruments, Inc. These "ats" files can only be created with version 3.12 or later of LAUD and they end with the file name extension "ats". They contain only driver Thiele-Small parameters. The LAUD import filter is available for both open back and sealed back drivers.

The Woofer Tester 2 import filter will import driver parameters from a "log" file from the WT2 measurement system by Smith & Larson Audio. These "log" files are created by selecting "Export to BassBox" in the WT2's "WooFile" menu. The Woofer Tester 2 import filter is available for both open back and sealed back drivers.

To import a driver, simply click on the "Import" button at the bottom of the Driver Properties window. An Import Driver T-S Parameters window like the one shown at the top of the next page will open to prompt you to choose a path and file name.

Using this import capability, you can design a box with BassBox Pro or Lite and then import it into X•over Pro to design the crossover network and view the overall response of the entire speaker in the graphs.

Note: The Import Driver T-S Parameters window will vary in appearance based on the version of Windows. For example, the "Files of type" list may not be labeled and will be located to the right of the "File name" input box in Windows 8, 7 and Vista.

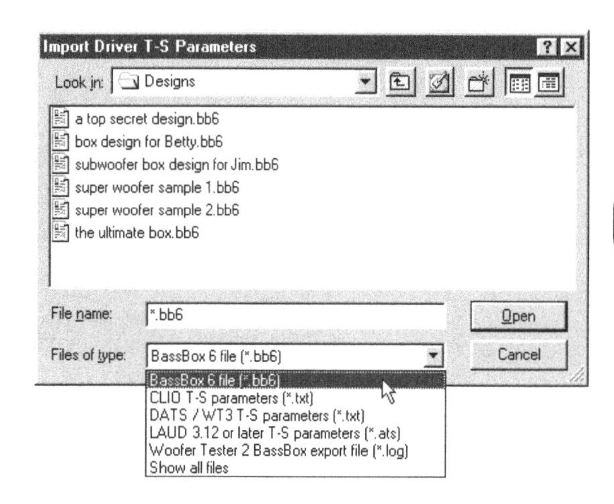

Manual Driver Data Entry

Driver information can be entered manually into X•over Pro by typing it directly into the input boxes in each of the tabs of the Driver Properties window. This provides an easy way to enter information directly from a driver manufacturer's spec sheet. You an also edit the driver information at any time. This includes driver information that was loaded from the driver database or imported from a BassBox Pro / Lite, CLIO or LAUD file.

The Driver Properties window is organized with six tabs as shown below:

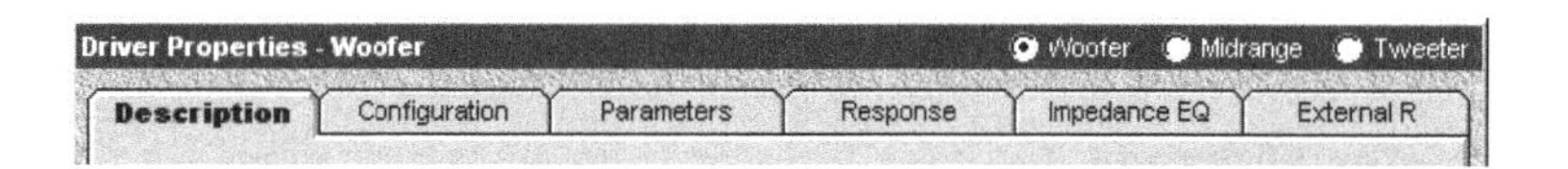

Remember that the information for only one driver will be displayed at a time. To view, enter or edit the woofer data, select "Woofer" in the title bar. Select "Midrange" or "Tweeter" for the midrange driver or tweeter, respectively.

The "Description", "Configuration", "Parameters" and "Response" tabs are concerned only with the drivers, themselves. The "Impedance EQ" tab can include the box information for open back drivers. This is because the box has a very significant effect on the impedance response of an open back driver and therefore on the design of an impedance equalization network for it. The "External R" tab includes information about the resistance of the speaker cables. There are separate resistance settings for the speaker cables between each crossover network filter and corresponding driver. There is only one set of settings for the resistance between the amplifier and the crossover network.

The remainder of this chapter will describe each tab.

2

Description

The "Description" tab contains optional descriptive information about a driver.

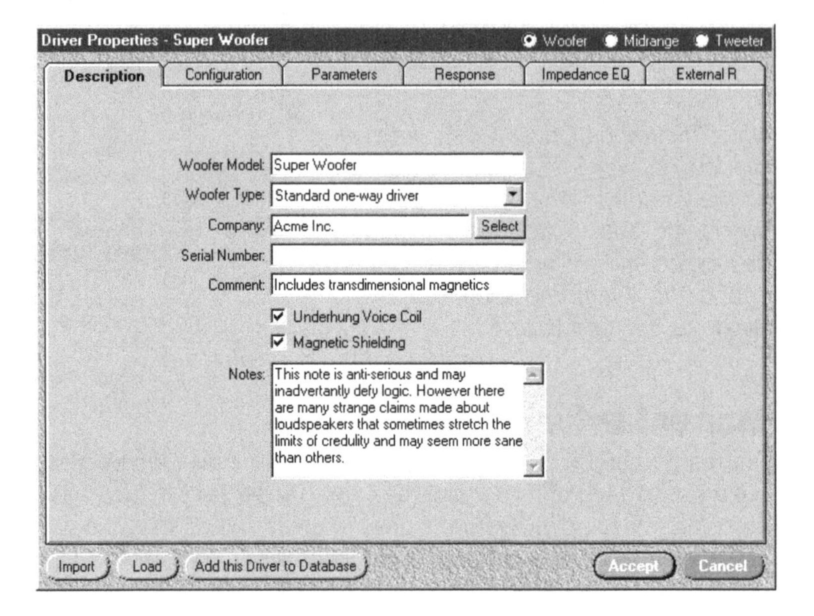

Woofer, Midrange, Tweeter Model The model name of the driver can be up to 30 characters long. Keep in mind that in the driver database a single manufacturer is not allowed to have more than one driver with the same model name.

Woofer, Midrange, Tweeter Type Four general driver types are provided: standard one-way drivers, two-way coaxial drivers, two-way coincident drivers and three-way drivers. Midrange drivers further divide the one-way drivers into open back and sealed back drivers. See Chapter 2 of the *Crossover Network Designer's Guide* earlier in the manual for more information.

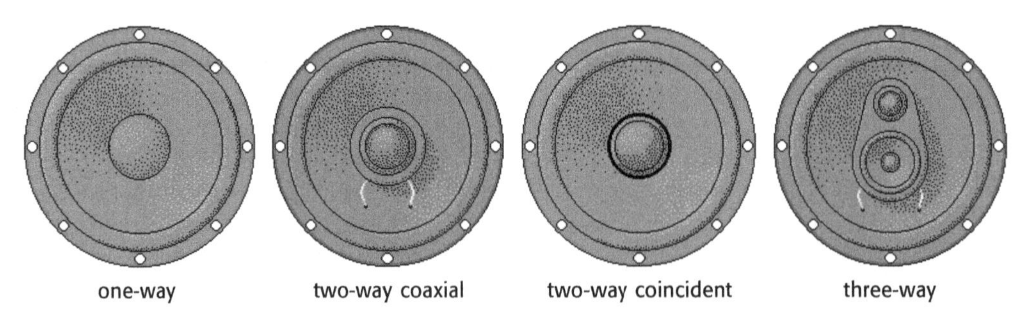

one-way two-way coaxial two-way coincident three-way

One-way drivers are the most common type of driver and contain only a single piston. Two-way and three-way drivers actually consist of more than one driver combined together in various ways.

Coaxial two-way drivers have an open back driver and tweeter. Both drivers are mounted like two wheels on an axle with the tweeter mounted in front. The magnet assemblies of the drivers are completely separate and the sound from the tweeter arrives ahead of the open back driver.

Coincident two-way drivers have an open back driver and a tweeter. But the tweeter is mounted in the acoustical center of the open back driver. The magnet assemblies are combined and the sound from the tweeter and the open back driver emanate at the same time and from the same apparent location. Because the sound from both drivers is "coincident" these drivers have superior time-domain characteristics, resulting in minimal phase cancellations in the crossover region.

Three-way drivers have a woofer, midrange driver and tweeter. The midrange driver and tweeter are usually mounted side by side in a common bracket. This assembly is then mounted in front of the woofer. Neither the midrange driver nor tweeter are coaxial or coincident with the woofer and the sound of all three driver elements will arrive at different times with the woofer lagging behind the other two.

Note: X•over Pro allows the larger driver in a two-way driver to be the "midrange" driver for the design so that it can be connected to the band-pass filter in a three-way crossover network. In this situation a separate woofer would be specified as the third driver that is connected to the low-pass filter of the network.

Two-way coaxial drivers and three-way drivers are commonly used in automotive applications where full-range sound from an extremely small space is desired. Two-way coincident drivers are not very common and are sometimes used in recording studio monitors of various types. They are also used in some higher-quality home hi-fi speaker systems.

Company The company name can be up to 40 characters in length and there are two ways to enter it. The preferred way is to use the "Select" button and choose a company name from the driver database. This insures that the spelling matches the spelling in the database in case the driver is added to the database later. The second way to enter the name is to type a name directly into the company input box.

Serial Number The serial number can be up to 30 characters long.

Comment The comment can be up to 50 characters long. It is a single-line note that will be included in the printouts.

Notes This is a multi-line note that is not limited in length and can contain more lengthy comments about the driver.

Underhung Voice Coil (open back drivers only) Check this checkbox when the driver has an underhung voice coil. Overhung voice coils are the most common type because they are typically less expensive to design and manufacture. With an overhung voice coil the height of the voice coil is greater than the height of the gap of the magnet. An underhung voice coil is just the opposite. The height of its voice coil is less than the height of the gap of the magnet so the entire voice always stays in the gap up to the maximum linear excursion (Xmax). This usually gives underhung voice coils an advantage in terms of linearity. But it can make them expensive when a large excursion is needed.

Ferrofluid (sealed back drivers only) Check this checkbox when the driver contains ferrofluid in the gap of the magnet. Ferrofluid is usually used to cool the voice coil, thereby increasing the power handling. It also damps the resonance, creating a much smoother impedance response.

Magnetic Shielding Typical drivers cannot be placed near a television, computer monitor or video monitor because the magnetic field of the magnet assembly of the driver will bend the path of the electrons in the picture tube causing the picture to distort. To avoid this problem, many drivers are now available with magnetic shielding. Check this checkbox if the driver has magnetic shielding.

Note: The driver description does not have to be entered in order to complete a crossover network design. However, it can make a design much more understandable to others.

Configuration

The "Configuration" tab can be ignored if only one of the selected drivers will be used in the design. If more than one is used, use this tab to set their mechanical and electrical configurations. This will also cause "net" parameter values to appear on the "Parameters" tab.

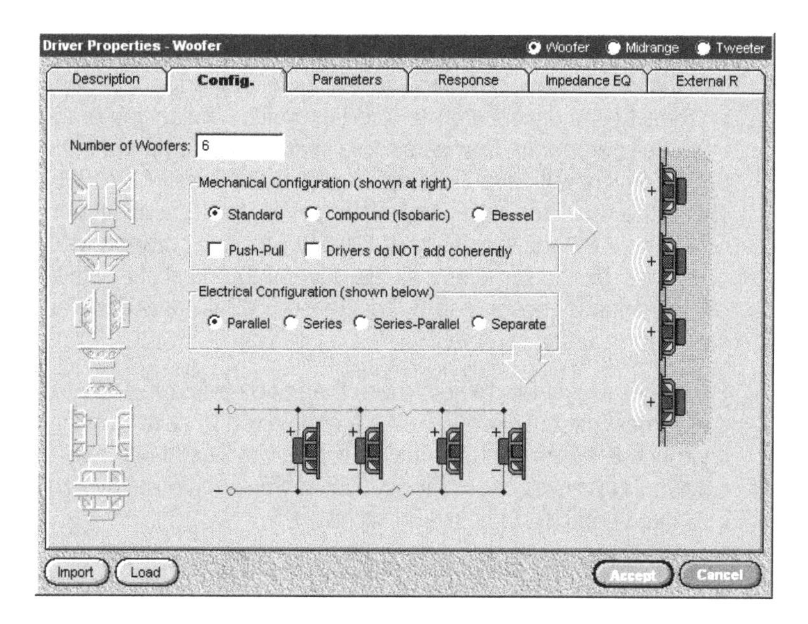

Number of Drivers X•over Pro allows you to specify any number of identical woofers, midrange drivers or tweeters for a design with one exception. The Bessel configuration requires either 5 or 6 drivers. *Note: Although Bessel arrays with more than 6 elements are possible, X•over Pro models only 5 and 6 element Bessel arrays.*

Mechanical Configuration Refers to the way the drivers are mounted in the box. Four choices are provided. The "standard" configuration should be chosen for all sealed back drivers.

Standard

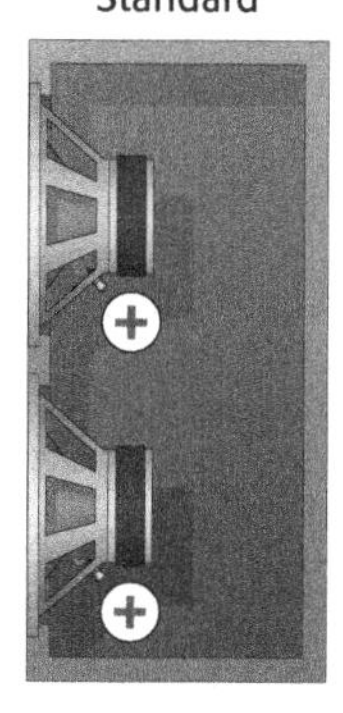

Standard This is the simplest way to mount multiple drivers. When the Push-Pull parameter is turned off, the drivers are all mounted facing the same direction and are wired with the same polarity. The box size must increase with each additional driver. Every doubling of drivers causes the 1-watt sensitivity (1-W SPL) to increase 3 dB if the parallel electrical configuration is chosen and decrease 3 dB if the series electrical configuration is chosen. The 2.83-volt sensitivity (2.8-V SPL) will increase 6 dB in parallel and remain unchanged in series.

2

Compound (Isobaric) This configuration mounts pairs of open back drivers closely together with a small isobaric (constant pressure) sealed chamber between them.

Compound

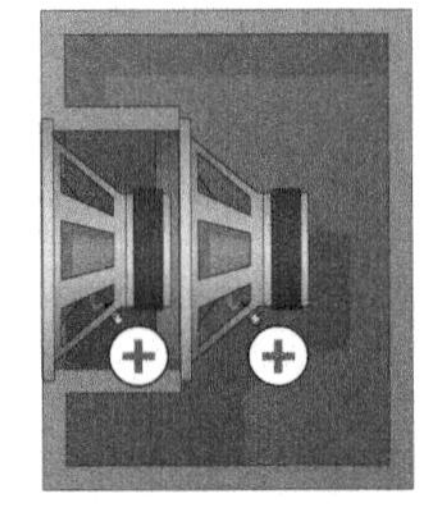

As such, there must be an even number of drivers. The most unique feature and the greatest advantage of this configuration is that the required box volume for a pair of compound drivers is half that of a single open back driver. The wiring polarity depends on the way the drivers are mounted as well as the setting of the Push-Pull parameter. Compared to a single driver, a pair of compound drivers will have 3 dB less sensitivity at 1 watt (1-W SPL). This is the same for both parallel and series electrical configurations. The 2.83-volt sensitivity (2.8-V SPL) for two compound drivers will be the same as a single driver if the parallel electrical configuration is chosen and will decrease 6 dB if the series electrical configuration is chosen.

Compound

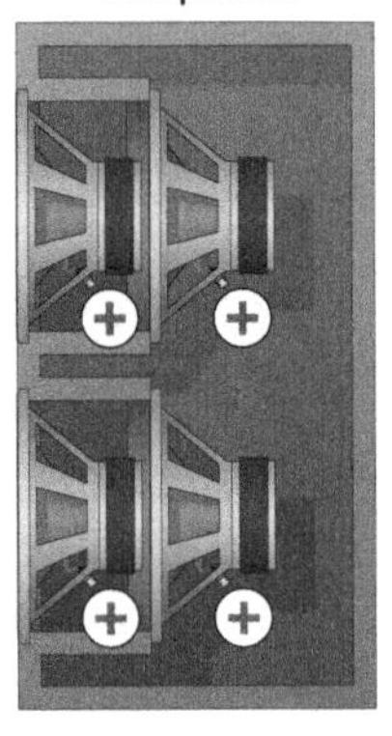

How does X•over Pro handle more than two compound drivers? It assumes that only two drivers are connected to each isobaric chamber. For example, four drivers would be divided into two compound pairs as shown at right. Six drivers would be divided into three compound pairs and so on.

Push-Pull The standard and compound mechanical configurations each have a push-pull option. It is only available when there are an even number of drivers (2, 4, 6, 8, etc.). When the push-pull option is used, every other driver is mounted in a reverse direction and is wired with opposite polarity. Examples are shown below:

Standard Push-Pull

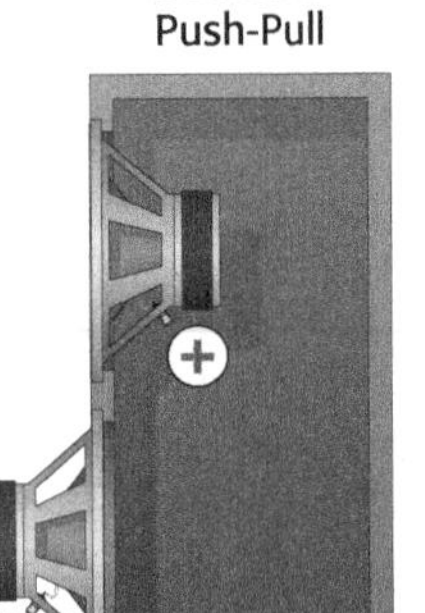

Compound Push-Pull

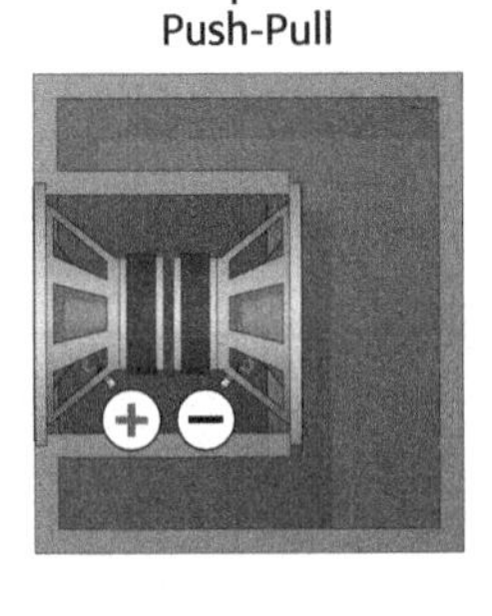

Compound Push-Pull

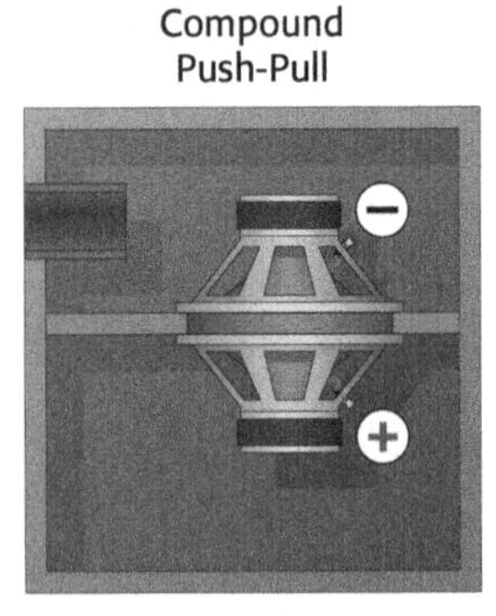

There are two different ways to mount a pair of compound push-pull drivers. One way (the middle illustration on the previous page) mounts them in an isobaric chamber with each driver pointing away from the other. The second way (the right illustration on the previous page) mounts them facing each other on a common wall. Of course, there is a hole in the wall so that the space between the two cones forms the isobaric chamber. The second method is popular with compound bandpass box designs.

Why use the push-pull option? Because it is a very effective way to reduce some kinds of distortion. Since the pistons of half of the drivers in a push-pull configuration move "in" while the other half move "out", many nonlinearities are cancelled. The result is a dramatic reduction in even-order distortion. *Note: This technique only works for open back drivers since sealed back drivers do not generate sound waves from the rear.*

Bessel This is a specialized mounting and wiring method that is available only when there are 5 or 6 drivers and only for closed box designs. The drivers are mounted in a line array and should be mounted as close together as possible with one important exception—a Bessel array with 6 drivers must leave a single blank space in the middle that is the size of one driver.

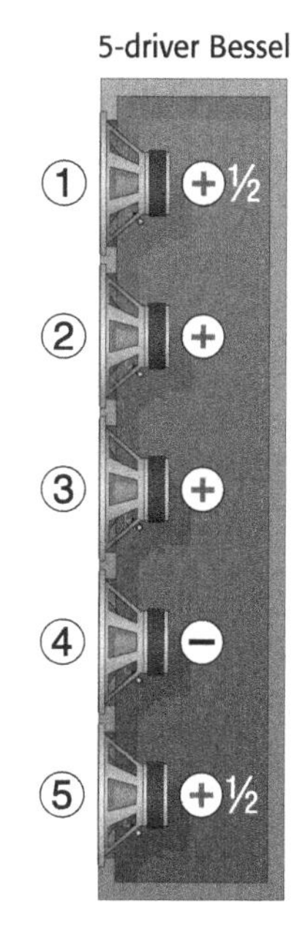

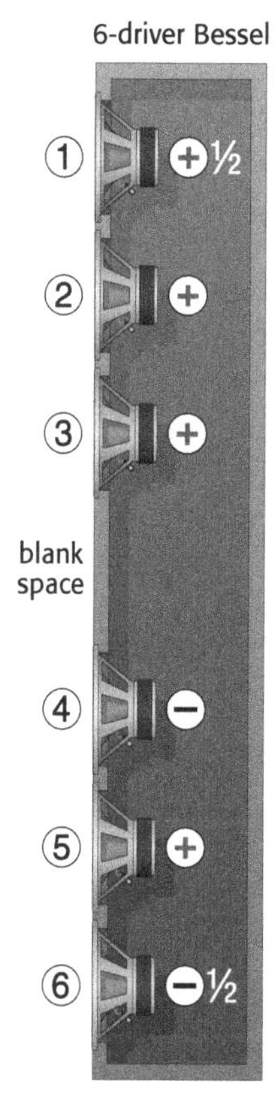

When a speaker has multiple drivers it is normal for there to be phase cancellations which cause the coverage pattern and sound quality to suffer. The unique feature of a Bessel array is that the phase cancellations are mostly eliminated, resulting in a combined coverage pattern that is nearly identical to a single driver. However, the pattern needs some distance to combine. One "rule of thumb" is to use a working distance that is 20 times the length of the array. This assumes the length of the array is measured from the center of the two outside drivers.

The real secret of a Bessel array is in its unique combination of mounting and wiring. A unique series-parallel method of wiring is required which controls the polarity and level of each driver so that their signals combine properly.

2

Drivers do NOT add coherently When multiple woofers, midrange drivers or tweeters are used, the sound waves emanating from them will combine to create a composite sound wave that is louder. (**Important:** This is NOT referring to the way a woofer will sum with a tweeter in a two-way speaker. Rather it is referring to the way two identical drivers will sum.) X•over Pro assumes that the drivers are all the same kind and, in most cases, that they are driven with identical signals so that their sound waves will usually sum coherently. This means that the net sound level will increase 6 dB with every doubling of drivers.

However, there may be occasions when the sound waves do not add coherently and the "Drivers do NOT add coherently" option should be turned on. In these cases the net sound level will increase only 3 dB for every doubling of drivers. The following list describes situations when coherent additions will not happen. Turn on the "Drivers do NOT add coherently" option for these situations:

- The drivers will not sum coherently if they are wired separately and are driven with different signals. For example, two woofers are mounted in a common cabinet but one is driven from a left stereo signal and the other is driven from a right stereo channel.
- The drivers will not sum coherently if they are mounted too far apart. Their center-to-center spacing should be no greater than one quarter (1/4) wavelength for the frequencies in their passband. This is usually not a problem for subwoofers because they are driven only with low-frequencies having long wavelengths. For example, many subwoofers use a crossover frequency of 100 Hz or less–the wavelength of 100 Hz is 136 inches (345 cm). As long as the drivers reproduce frequencies that are not higher than 100 Hz then the drivers in the subwoofer can be mounted as far as 34 inches (86 cm) apart because this is one quarter of 136 inches.
- Coherent summing is not relevant to compound configurations because only one of each isobaric pair of drivers will radiate sound outside the box. The "Drivers do NOT add coherently" feature will be disabled when a "Compound" mechanical configuration is chosen.

The drivers are always assumed to sum coherently when a Bessel array is chosen.

Electrical Configuration Describes the driver wiring. Four choices are provided.

Parallel This is usually the best electrical configuration because it offers the best electrical isolation between drivers. However, it has one serious drawback: if too many drivers are used, the net impedance seen by the amplifier can become extremely low and, if the amplifier does not have sufficient current headroom, it may blow a fuse, activate its protective circuitry or fail if it has no protective circuitry. For example, four 8 ohm drivers wired in parallel have a net impedance of 2 ohms. Four 4 ohm drivers wired in parallel have a net impedance of only 1 ohm! Two examples of parallel wiring are shown below:

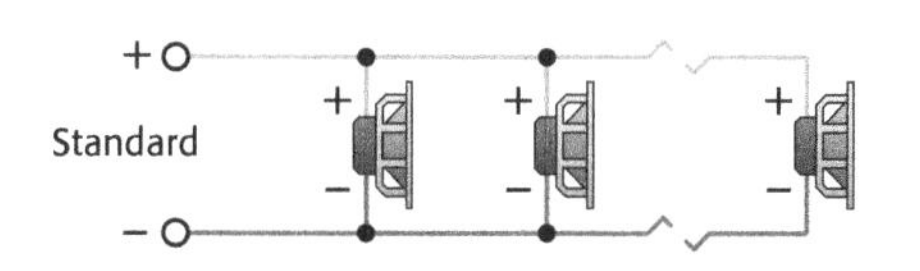

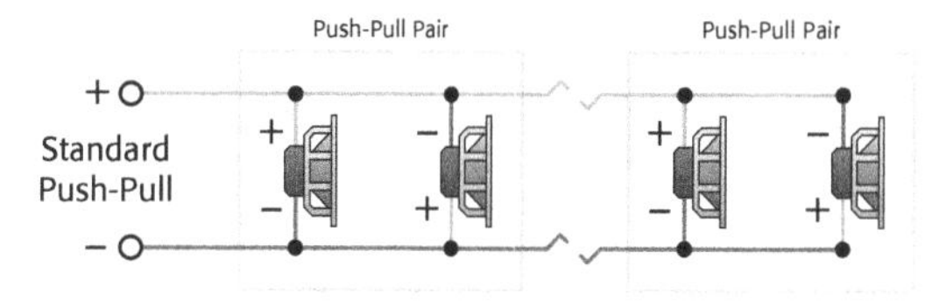

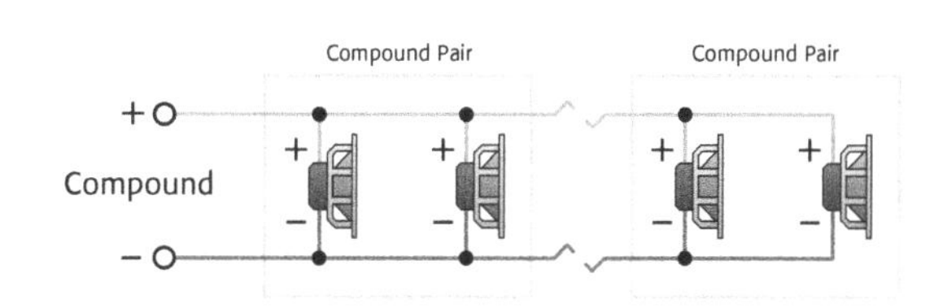

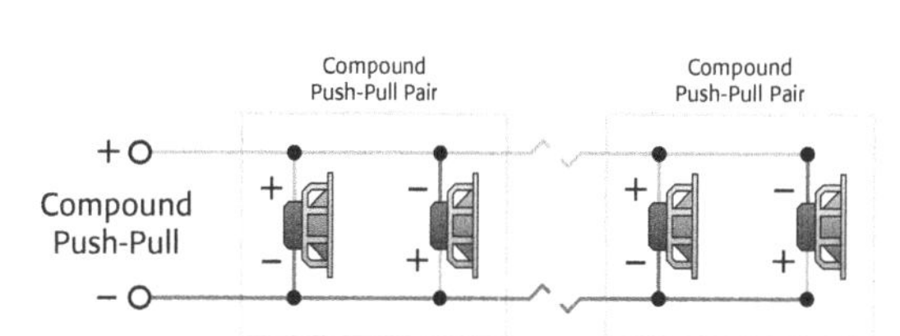

Notice that the polarity of every other driver is reversed in the push-pull examples.

Series This is usually the least desirable electrical configuration because the drivers have the greatest interaction. An "open" circuit anywhere in the wiring (such as a blown driver) will prevent sound from reaching all of the remaining drivers. However, it has one advantage: the net impedance seen by the amplifier increases with each additional driver. For example, two 8 ohm drivers wired in series have a net impedance of 16 ohms. Four 8 ohm drivers wired in series have a net impedance of 32 ohms. Be aware that as the impedance rises, the drivers draw less power from the amplifier. Two examples of series wiring are shown at the top of the next page:

2

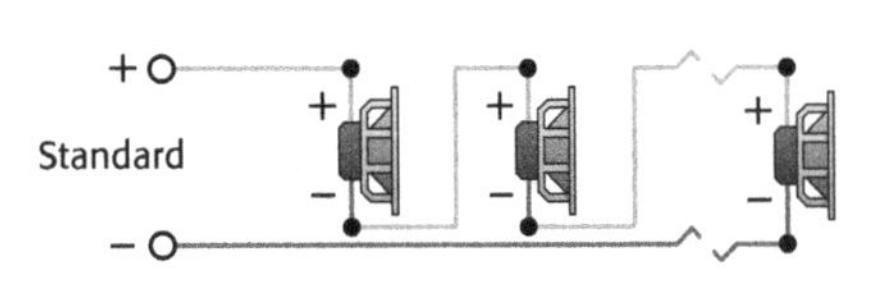

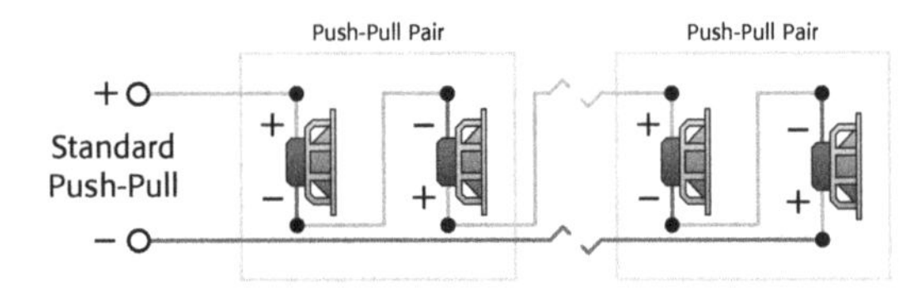

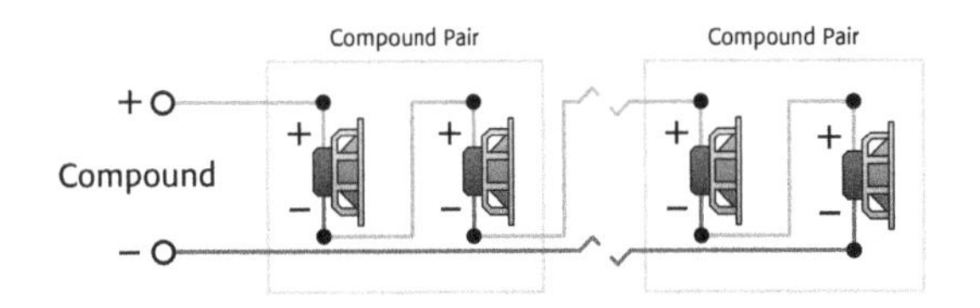

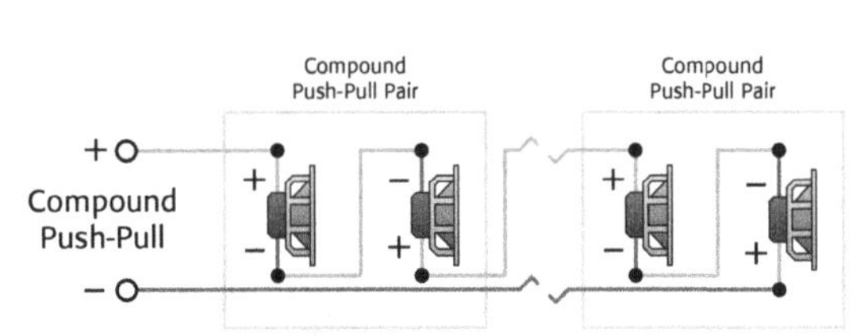

Series-Parallel This is a combination of both series and parallel wiring which is sometimes used to prevent the impedance from going too low or too high. X•over Pro assumes that the drivers are paired in series and that each pair is in turn wired in parallel with all the other pairs (except for Bessel configurations). As a result, there must be a minimum of four drivers and the number of drivers must be even to use this configuration. Four examples of series-parallel wiring are shown below:

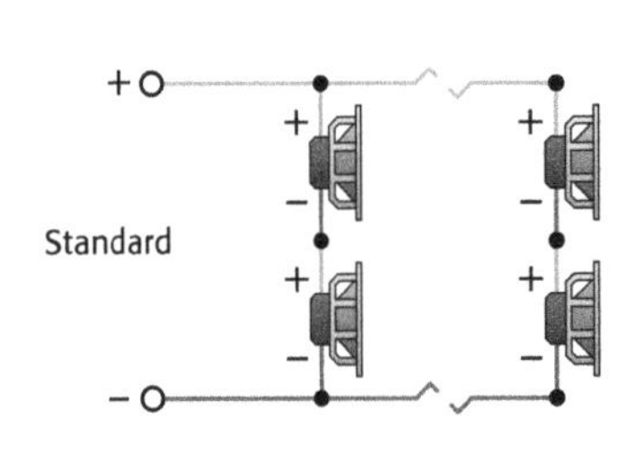

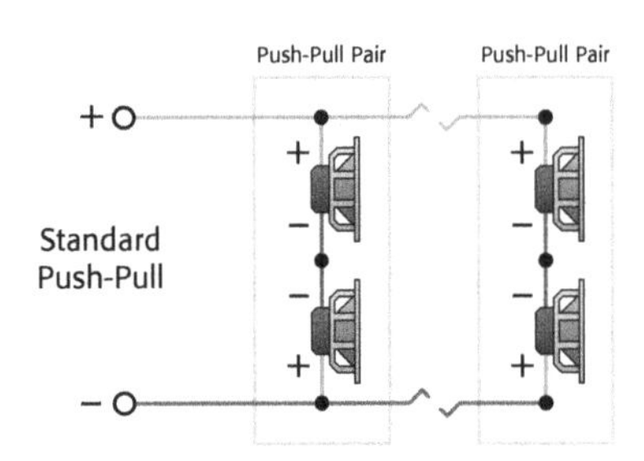

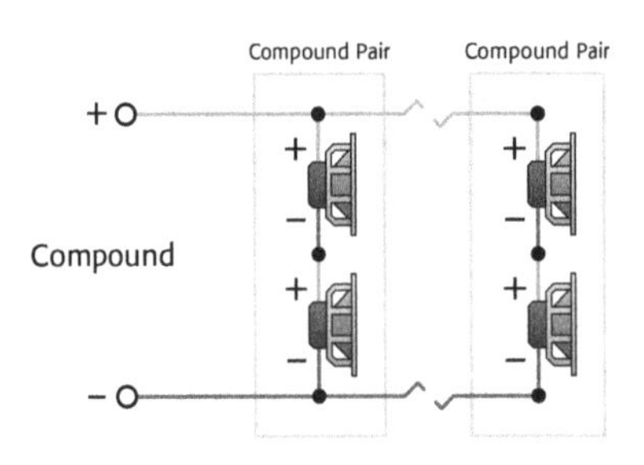

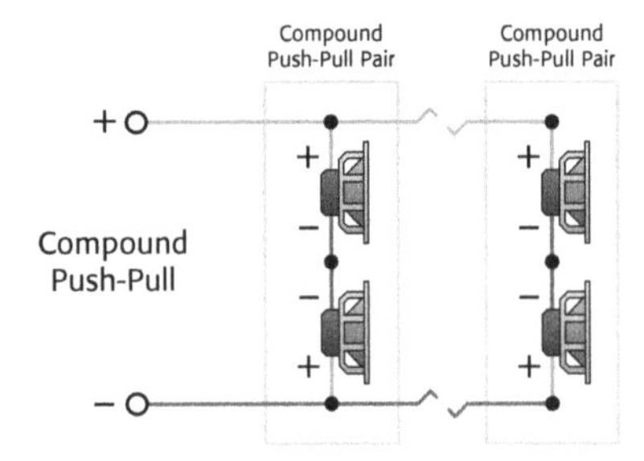

Bessel designs require special series-parallel wiring in order to control the level and polarity of each driver. The wiring for 5-driver and 6-driver Bessel configurations are shown below:

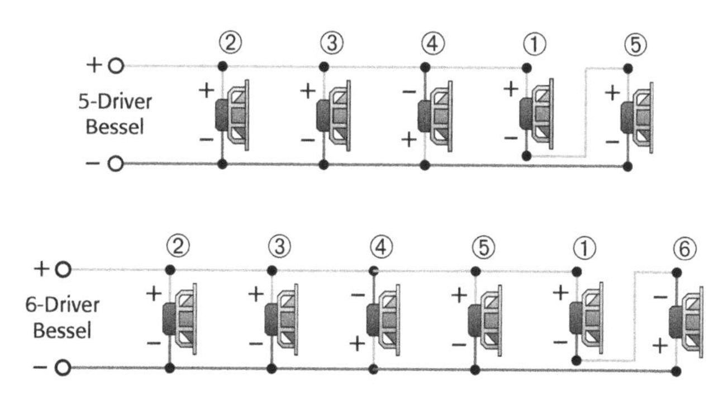

Separate Select this configuration if the drivers will not be connected together but will instead be each driven by a separate amplifier channel with a separate filter. For example, a two-way speaker with four woofers will need four identical low-pass filters (one for each woofer) when the woofers are wired separately.

When the "Separate" option is first selected, the "Drivers do NOT add coherently" option will also be turned on since the program has no way of knowing if the drivers will be driven with the same signal. If they are and their sound waves will sum coherently, then the "Drivers do NOT add coherently" option should be turned off. *Note: When the "Separate" configuration is selected, the "Drivers do NOT add coherently" option will determine whether some of the graphs show the net response or the response of just one driver. See the descriptions of each graph for more details.*

Please do not confuse the "Separate" electrical configuration setting of the "Configuration" tab with the "Separate Filters" command of the Network menu. The filters can be combined in a 2-way or 3-way crossover network and still have multiple drivers with separate wiring. As explained above, separately wired drivers will require separate filters. For example, a 2-way crossover network with a single tweeter and two separately wired woofers will require one high-pass filter for the tweeter and two identical low-pass filters, one for each of the woofers. *Note: The schematic and component parts list will include just one each of the low-pass filters.*

2

Parameters

The "Parameters" tab contains the Thiele-Small and electromechanical parameters for a driver. "Thiele-Small", often abbreviated "T-S", refers to Neville Thiele and Richard Small who first popularized modern methods of speaker modeling.

Mode X•over Pro supports a total of 21 parameters but not all of them are required at the same time. For this reason, two modes are provided for the "Parameters" tab, the "normal" mode and the "expert" mode. They are selected by turning on and off the "Expert Mode" checkbox in the upper right corner of the "Parameters" tab as shown in the illustration below. The default mode can be set in the "Driver" tab of the Preferences window. Let's examine each mode in detail:

Normal Mode (The "Expert Mode" checkbox is turned OFF.) The "normal mode" (shown below) presents a less complex version of the "Parameters" tab with only 7-9 frequently used driver parameters.

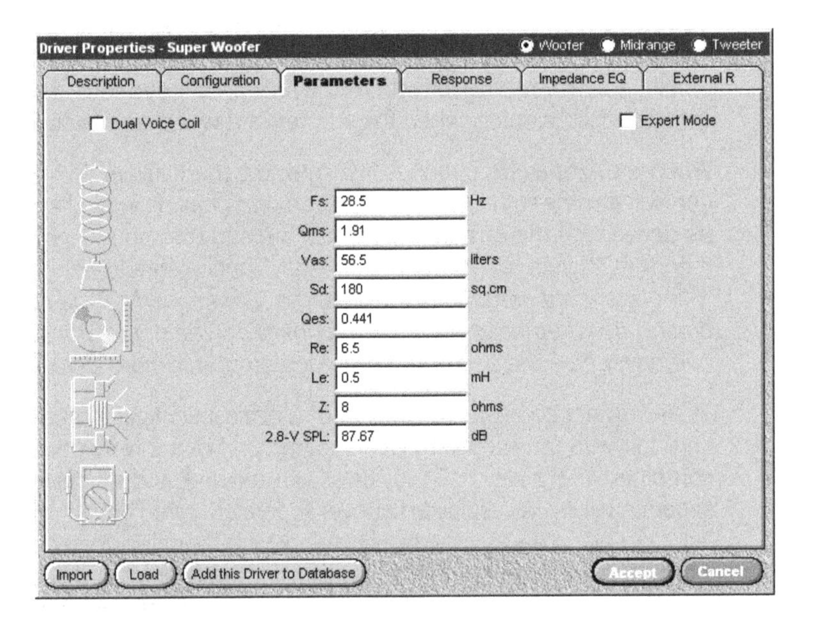

The minimum parameters for a full design are included. These are the parameters required to design an impedance equalization network for a driver and enable the graphs to display the response of the filters and the system. The minimum parameters vary slightly for open back and sealed back drivers. Examples are shown on the next page:

Open Back Drivers	Sealed Back Drivers
Fs	Fs
Qms	Qms
Vas	–
Sd	–
Qes	Qes
Re	Re
Le	Le
Z	Z
2.8-V SPL	2.8-V SPL

It is possible to use less parameters but this will diminish the capabilities of X•over Pro. If necessary, a crossover network can be designed with just the DC resistance (Re) or impedance (Z) of each driver. However, since the crossover network wants to see a predominantly resistive load and the speaker drivers are actually reactive loads, the crossover network will vary more or less from the expected results. The variance can be lessened by keeping the crossover frequencies as far away from the driver resonance frequencies as possible (at least one octave but preferably more).

Some of the minimum parameters can be swapped with others that are not visible. For example, Qms and Qes can be swapped with Qts and BL *(note: BL is not needed for sealed back drivers)*. Sd can be swapped with Dia. The 2.8-V SPL can be swapped with ηo or the 1-W SPL. See Chapter 11 for information on setting default parameters.

To switch one of these parameters, simply click on its label as shown below:

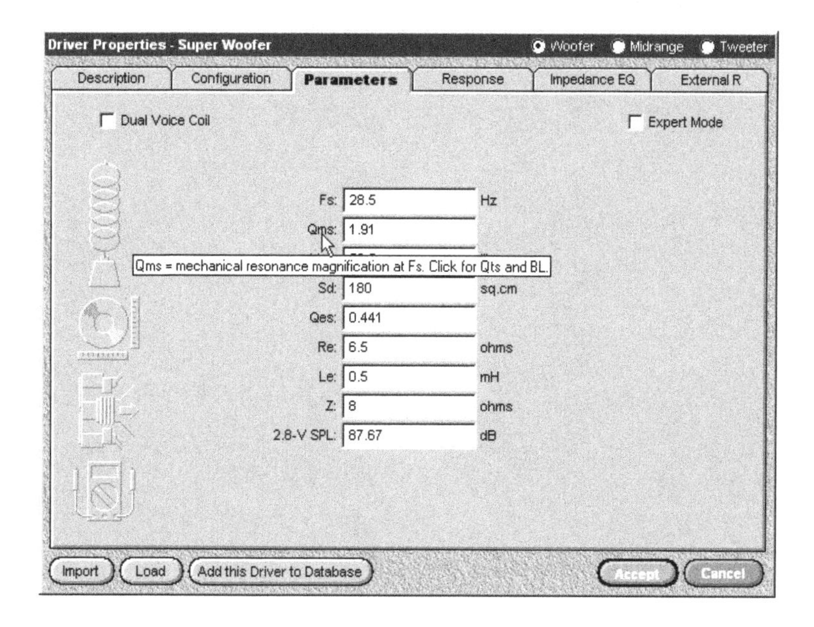

2

The illustration below shows some of the alternate parameters.

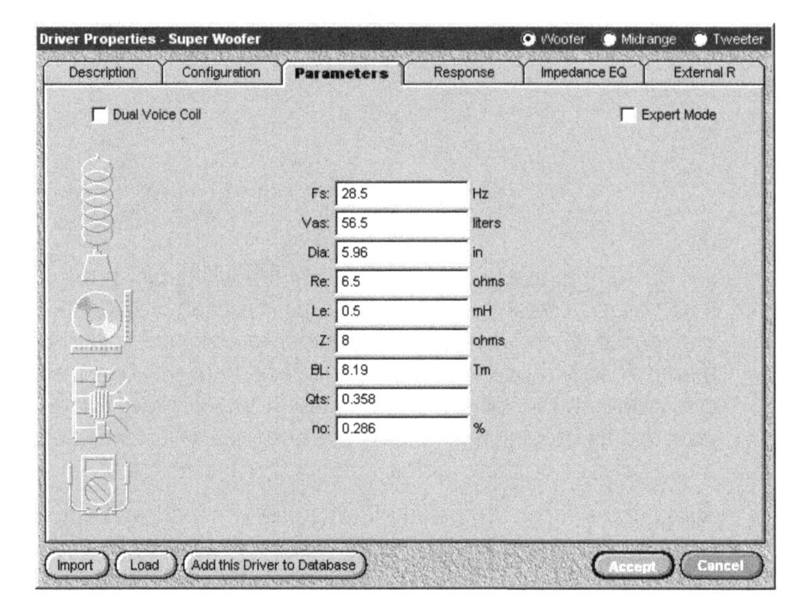

The selectable default parameters for the normal mode (Qms, Qes, Qts, BL, Sd, Dia, ηo, 2.8-V SPL, 1-W SPL) can be set with the "Driver" tab of the Preferences window (see Chapter 11 for more information).

Expert Mode (The "Expert Mode" checkbox is turned ON.) The "expert mode" presents a complete set of 21 Thiele-Small and electromechanical parameters. They are grouped according to their type (Mechanical, Electrical and Electromechanical).

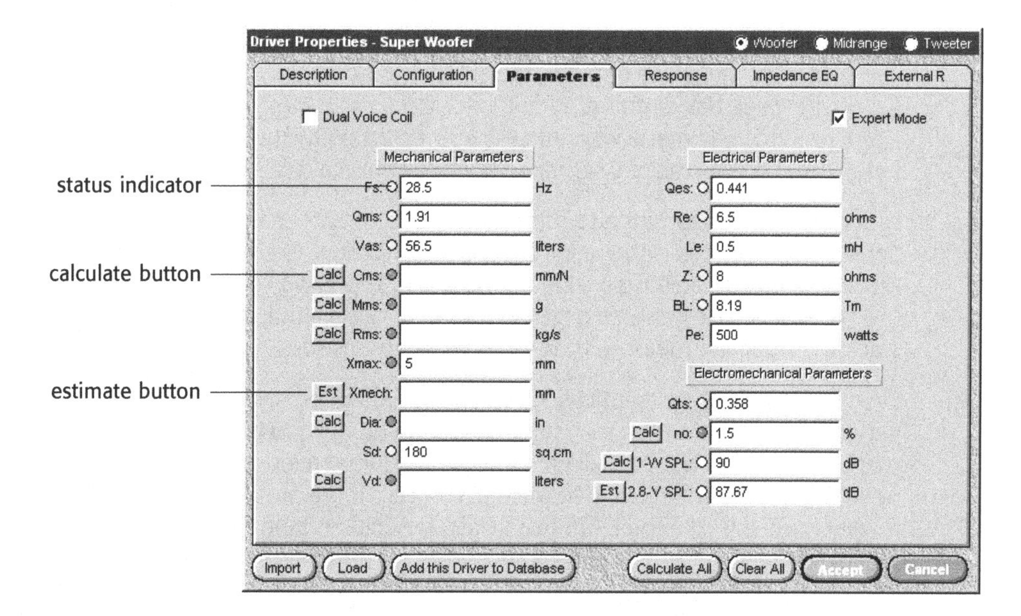

A key feature of the "expert mode" is its automatic tolerance testing of many parameters. A status indicator shows whether a parameter passed or failed. The tolerance settings for the tests can be set with the "Driver" tab of the Preferences window. This self-analyzing feature can quickly reveal "typos" and other mistakes immediately when they are made. It also provides a convenient way to see if a manufacturer's data contains mistakes or has been "fudged."

Note: Three parameters, Xmech, Le and Pe, do not have status indicators because they cannot be verified.

The color code for the parameter status indicators is:

Grey – there is not enough information available to test the parameter.
Green – the parameter is probably correct.
Yellow – the parameter appears to be slightly out of tolerance.
Red – the parameter is probably incorrect.

When the status indicators of multiple parameters turn red it does not mean that they are all necessarily incorrect. When two or more related parameters fail a test, X•over Pro can't know which one is incorrect and so it must flag them all. The default tolerance values are ±3% for yellow and ±7% for red.

What are some of the reasons that cause a parameter value to be out of tolerance?

- The wrong units were selected before the parameter was entered.
- The decimal place is incorrect.
- The value has been rounded too much.
- An error was made when the driver was tested by the manufacturer.
- An error was made when the parameter was calculated by the manufacturer.

Unfortunately, it is common to find one or more "bogus" parameters in driver specifications. Most often this is the result of "honest" mistakes. Even some of the drivers in the X•over Pro driver database are out of tolerance. This is why critical speaker designers use driver databases only as a starting point for design and, after obtaining some sample drivers, will measure their parameters, themselves..

Even if you are not doing a "critical" design, there will be times when you simply can't get good data. The drivers may no longer be in production or, as strange as it sounds, a manufacturer may not be willing to release the information. In such cases you, too, will need to measure the driver parameters, yourself. There are inexpensive devices on the market to do this, like the DATS from Dayton Audio and the Woofer Tester 2 from Smith & Larson.

Of course, you can always choose a driver from a manufacturer who is able and willing to provide accurate and complete data about their products.

Calc, Est Whenever a parameter is blank or out of tolerance, the "expert mode" will check to see if there is enough information available to calculate or estimate its value. If there is, a "Calc" or "Est" button will appear to the left of the parameter label. Note that special circumstances exist for the Qms, Qes and Le parameters.

A "Calc" button will appear beside the Qms and/or Qes parameters whenever their values can be calculated from the other parameters. However, if either Qms or Qes is blank and there are not enough known parameters to calculate them, then an "Est" button will appear. Clicking it will cause the Driver Q Estimator window shown on the next page to open.

This window enables you to estimate Qms and/or Qes from the shape of the driver's free air resonance peak from the impedance graph supplied by the driver's manufacturer. Follow the on-screen instructions to estimate one or both of the Q parameters. *Note: The total Q (Qts) that results from Qms and Qes will also be calculated and displayed.*

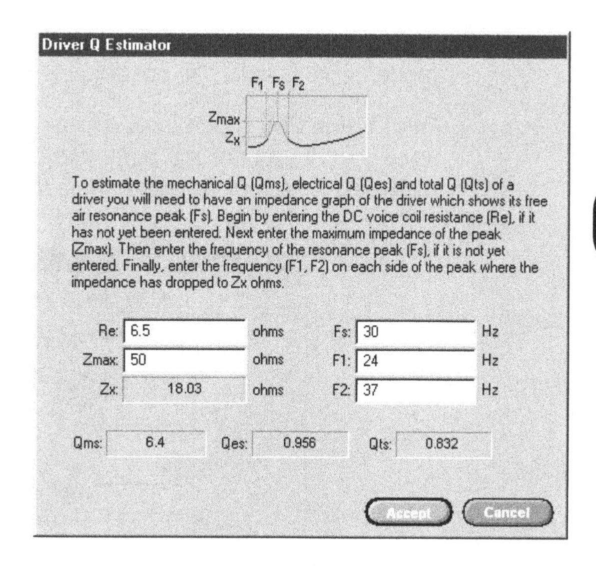

The Le parameter cannot be calculated from the other parameters. So, when Le is blank, an "Est" button will appear. Clicking on it opens the Driver Inductance Estimator window shown at right.

This window enables you to estimate Le from the shape of the high-frequency impedance rise from the impedance graph supplied by the driver's manufacturer. Follow the on-screen instructions to estimate Le.

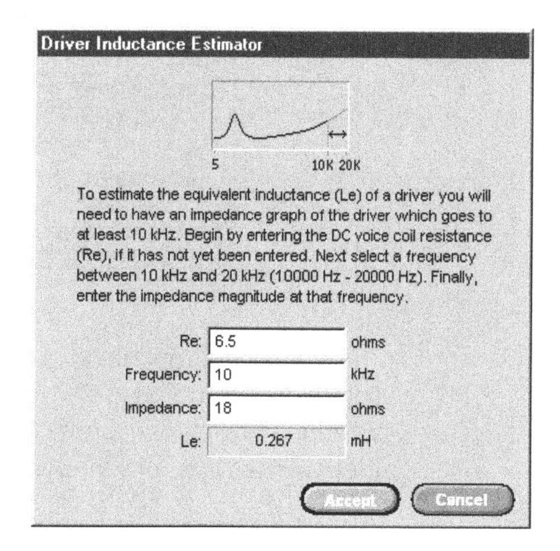

Calculate All, Clear All A "Calculate All" and "Clear All" button (shown below) are available below the "Parameters" tab when the "expert mode" is selected.

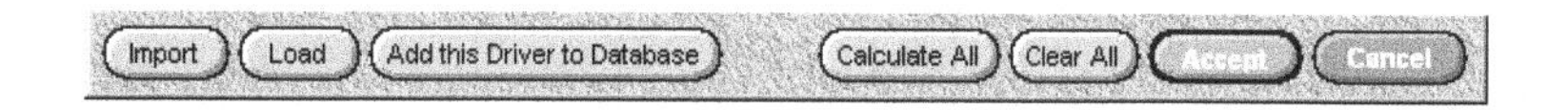

These buttons do just what their names imply, calculate all possible unknown parameter values (except Le and Pe) or clear all parameter values.

The remainder of this topic on the "Parameters" tab will describe the inputs for the "expert mode" since it covers all possible inputs.

Dual Voice Coil (open back drivers only) Used to identify open back drivers with dual voice coils. It expands the parameters to include separate inputs for each voice coil wiring method. An example is shown below:

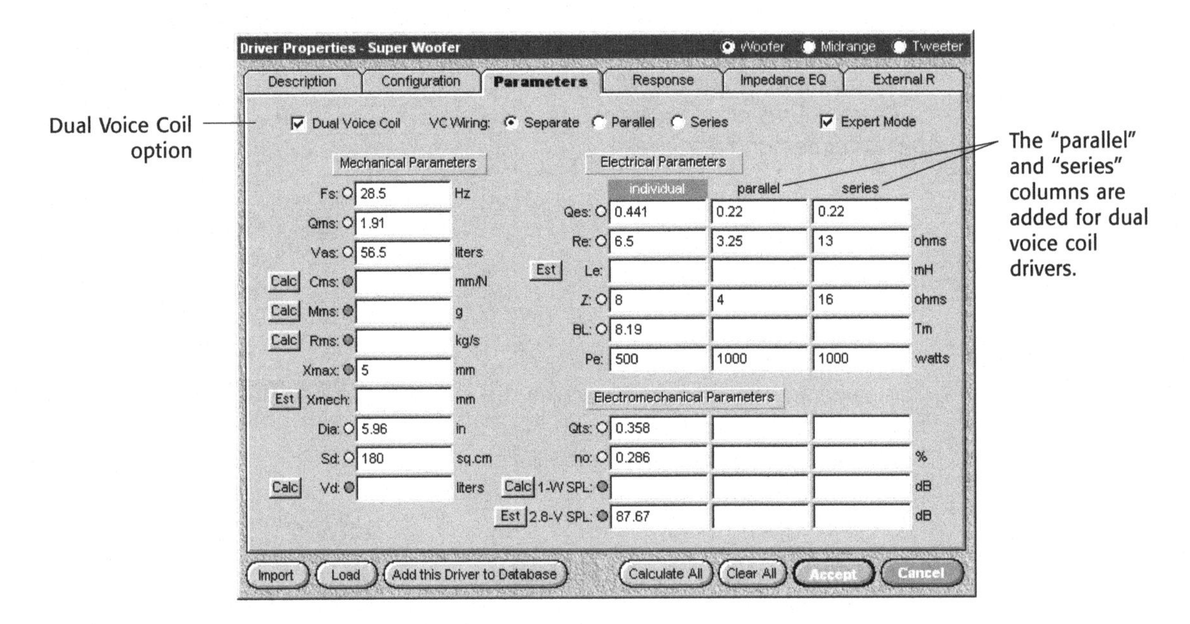

The "VC Wiring" option selects the wiring method. Select "Separate" if each voice coil will be connected to a separate amplifier channel and filter. Select "Parallel" if the voice coils will be wired in parallel and connected to the same amplifier channel and filter. Select "Series" if they will be wired in series and connected to the same amplifier channel and filter.

The electrical and electromechanical parameters will have a different value depending on the wiring method used for the dual voice coil. The mechanical parameters are not affected by the dual voice coil's wiring method. Some of the electrical and electromechanical parameters can be estimated if a value for at least one wiring method has been entered for the parameter. To estimate a value, double-click (🖱🖱) in the input box of the empty parameter. For example, double-clicking in the Re input box in the "parallel" column of the above illustration will result in an estimated value for Re of 3.25 ohms because the "parallel" value of Re is usually one half that of each separate (individual) voice coil.

The parameters for the selected wiring method will be used for all internal calculations. *Note: When the dual voice coil option is enabled, the status indicators of the electrical and electromechanical parameters will display only the status of the selected wiring method. And the "Calc" and "Est" buttons of the electrical and electromechanical parameters will only calculate the values of the selected wiring method.*

Mechanical Parameters

Fs The free air resonance frequency of the driver. Units: Hz.

Qms The mechanical resonance magnification of the driver at Fs.

Vas The volume of air equal to the mechanical compliance of the driver's suspension. Units: liters, cu.cm, cu.m, cu.ft or cu.in.

Cms The mechanical compliance of the driver's suspension. Units: μm/N, mm/N, cm/N, m/N or in/lb.

Mms The mass of the diaphragm or piston including the air load. Units: g, kg or oz.

Rms The mechanical resistance of the driver resulting from its suspension losses. Units: kg/s, lb/s or mohms.

Xmax The maximum linear excursion of the driver. It is measured in one direction from rest. Units: mm, cm, m or in. (P-P or peak-to-peak values must be halved.)

Xmech The maximum mechanical excursion of the driver. It is measured in one direction from rest: Units: mm, cm, m or in. (P-P or peak-to-peak values must be halved.)

Dia The diaphragm or piston diameter of the driver. It is usually measured from the middle of the surround as shown below. Units: mm, cm, m or in.

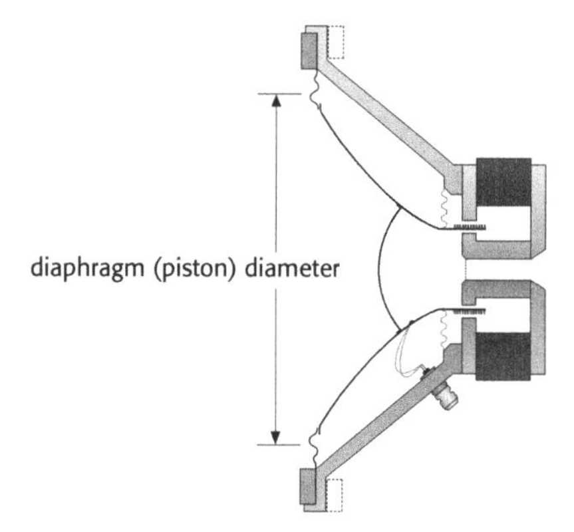

Sd The diaphragm or piston area of the driver. Units: sq.mm, sq.cm, sq.m or sq.in.

Vd The displacement volume of the diaphragm or piston at Xmax. Units: liters, cu.cm, cu.m, cu.ft or cu.in.

Electrical Parameters

Qes The electrical resonance magnification of the driver at Fs.

Re The DC voice coil resistance of the driver. Units: ohms.

Le An inductance value that is approximately equivalent to the upper frequency inductive reactance of the driver's voice coil. Units: mH.

Z The nominal impedance of the voice coil of the driver. Units: ohms.

BL The motor strength of the driver. Units: Tm, N/A, Tft or lb/A.

Pe The maximum electrical input power that the driver can handle. Units: watts.

Electromechanical Parameters

Qts The total resonance magnification of the driver at Fs.

ηo (eta zero) The half-space reference efficiency of the driver. Units: percent.

1-W SPL The sensitivity of the driver at 1 meter (3.3 feet) with 1 watt input. Units: dB.

2.8-V SPL The sensitivity of the driver at 1 meter (3.3 feet) with 2.83 volt input. Units: dB.

The units of many of the parameters can be easily changed. Simply click on the unit label as shown below. The units will advance each time their label is clicked.

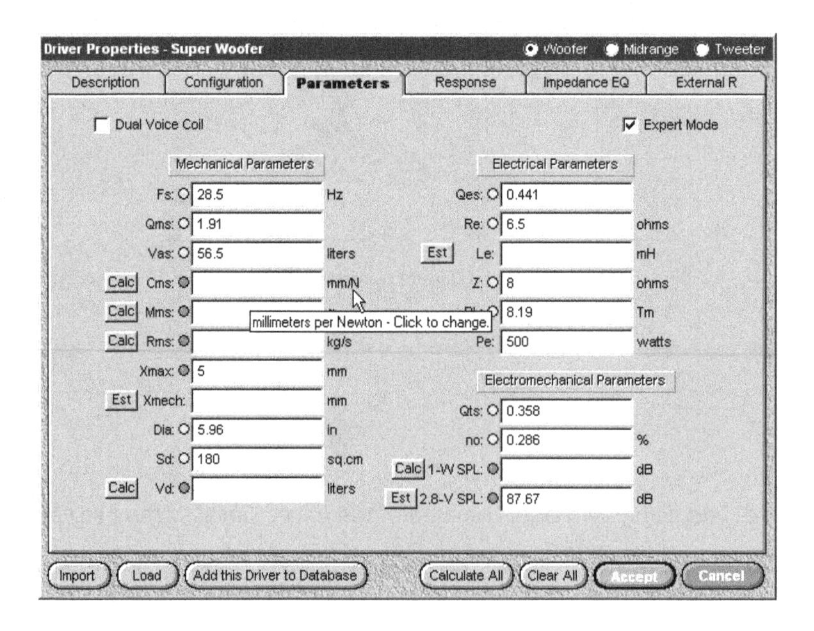

Response

The "Response" tab contains the normalized acoustic response of a driver (it must be normalized to the predicted response of the driver). It is very important that the acoustic data entered for the driver not include the response of a test box or baffle. The acoustic data should include only the +/– variations which occur when the measured response differs from the predicted Thiele-Small response.

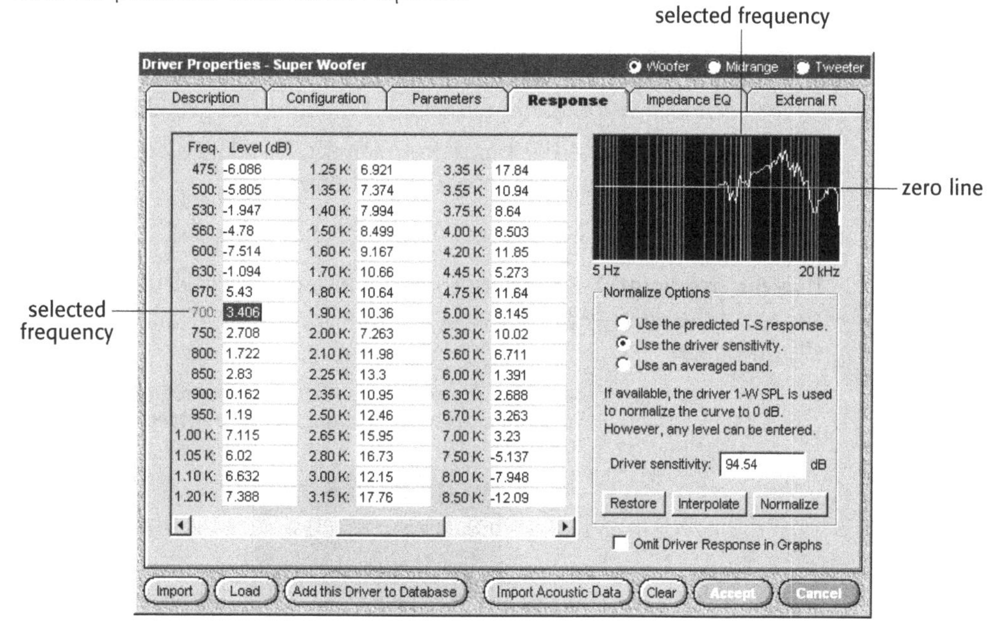

Why enter the driver's acoustic response? Because it improves the accuracy of the Normalized Amplitude Response graph by showing how acoustic anomalies will affect the design.

There are two ways to enter the acoustic response. It can be manually entered using the individual "Level" input boxes for each data point or it can be imported from one of several measurement systems.

The preview graph in the upper right corner shows the shape of the response while you edit or import data. Notice also that the preview graph includes a "zero line" to help you adjust the overall level of the data so that it is normalized. A red vertical line is also provided to show what part of the response will be changed if the input box of the selected frequency is changed.

Level The normalized amplitude response in dB for each acoustic data point. The response of the driver can be manually entered or edited with the "Level" input boxes. There are 134 data points from 5 Hz to 20 kHz. Use the horizontal scroll bar to access the "Level" input boxes for all of the data points. *Note: You do not have to enter a value for every acoustic data point. You can leave some blank and then use the "Interpolate" button later to estimate the missing values.*

Normalize Options The acoustical data must be normalized to the predicted response of the driver. (Zero dB represents the predicted response of the driver as defined by its Thiele-Small parameters and, if present, its box parameters.) This usually involves two-steps:

1. Adjust the overall level of the acoustic data so that the flat region of the response curve is level with the zero line in the preview graph. This can be accomplished with the second, third or fourth normalization options listed on the following pages.
2. Subtract the predicted response from the acoustic data. This can be accomplished with the first normalization option listed below.

Each normalization option is described next:

Use the predicted T-S response When this option is selected, the predicted response of the driver will be added to the preview graph with an orange plot line as shown below.

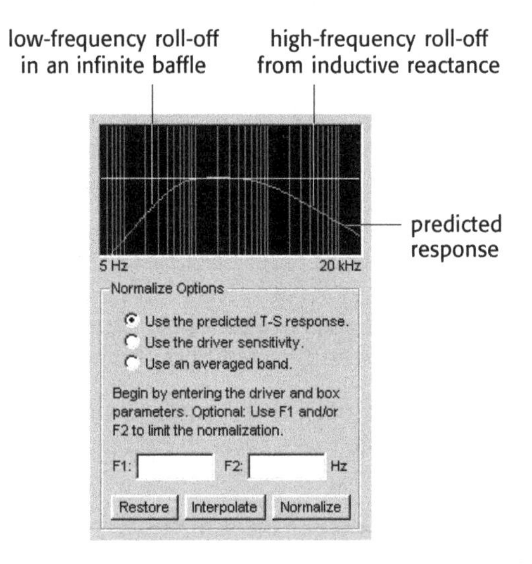

The predicted response should approximately match the shape of the acoustic response curve as shown in the left example in the illustration on the next page. If it does not, then one of two problems exits: 1) either there is an error in the driver or box parameters, or 2) there is a problem with the acoustic data.

The predicted response of open back drivers includes the box response if box parameters are entered. Otherwise it represents the driver in an infinite baffle. The predicted response can be changed by changing the driver parameters on the "Parameters" tab or by changing the box parameters on the "Impedance EQ" tab. **Important:** The box parameters should duplicate the test box used when the driver was measured. The box will be ignored for all sealed back drivers. *Note: An infinite baffle can be simulated by selecting a closed box and using a huge box volume (Vb) and huge QL value.* If necessary, remember to restore the driver and box parameters after normalizing the acoustic response.

When the "Normalize" button is clicked, the difference between the predicted response and the measured acoustic response will be calculated. This "difference" will be entered as the "normalized" acoustic response. The illustration below shows the same acoustic data before and after normalization:

BEFORE Normalization

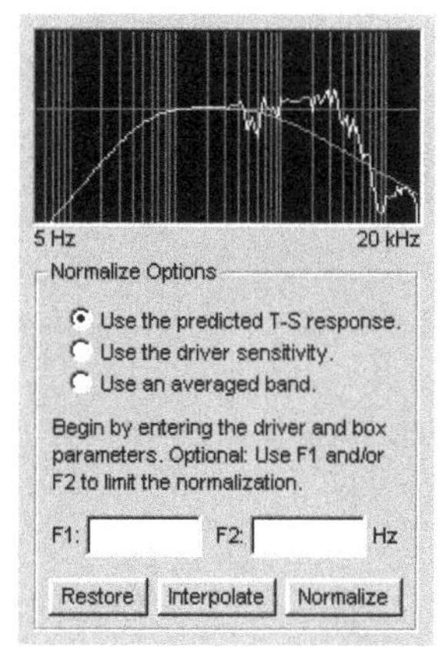

AFTER Normalization

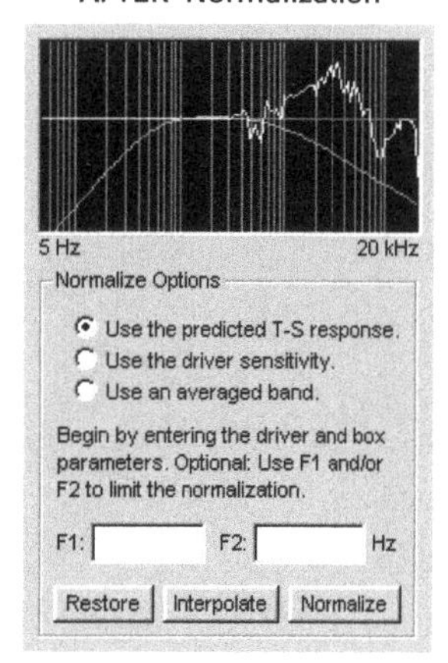

If desired, this method of normalization can be limited with F1 and F2. Use F1 to prevent low-frequency data from being normalized. Use F2 to prevent high-frequency data from being normalized. For example, enter "500" into F1 if you do NOT want acoustic data below 500 Hz to be normalized. Enter "7000" into F2 if you do NOT want acoustic data above 7 kHz to be normalized. Either or both F1 and F2 can be left blank if you do not want to limit normalization.

Tip: Do not leave the "Use the predicted T-S response" option selected after you have finished normalizing the acoustic data because it can slow the operation of the program on some computers. This is because X•over Pro recalculates the predicted response every time a change is made to a driver, box or circuit parameter. Instead, select either the second or third normalization option before you leave the "Response" tab after you have finished preparing the acoustic data.

Use the driver sensitivity This normalization option does not change the shape of the acoustic response like the first option. Instead, it adjusts the overall level of the acoustic data. For example, an acoustic measurement may represent a 1-watt, 1-meter sound pressure level. In cases like this, the acoustic response may not be visible in the preview graph because the acoustic level is too high. For example, the flat portion of the response may be between 85 and 95 dB. Use this normalization option to adjust the level of the acoustic data so that it is level with the zero line in the preview graph. This is best done before normalizing the data to the predicted response.

Depending on the nature of the acoustic data, it can sometimes be normalized to the zero line with the driver sensitivity rating. However, any level can be used—even a negative level if the acoustic data is below the zero line. Let's examine an arbitrary example:

BEFORE Normalization

5 Hz 20 kHz
Normalize Options
Use the predicted T-S response.
Use the driver sensitivity.
Use an averaged band.
If available, the driver 1-W SPL is used to normalize the curve to 0 dB. However, any level can be entered.
Driver sensitivity: 94.54 dB
Restore | Interpolate | Normalize

AFTER Normalization

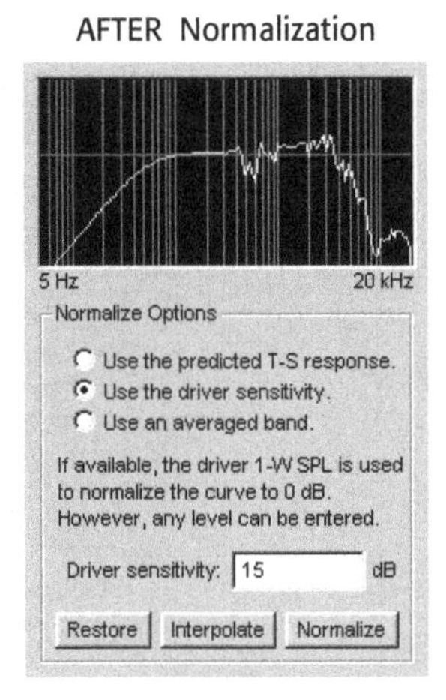

In the above example, the flat region of the acoustic data is about 15 dB above the zero line as shown on the left. To correct this, replace the "Driver sensitivity" value with "15" and click the "Normalize" button to subtract 15 from each of the acoustic data points. The result is shown above in the example on the right.

Use an averaged band This normalization option does not change the shape of the acoustic response like the first option. Instead, it adjusts the overall level of the acoustic data like the second option. It does this by subtracting the average level of a selected frequency band from the acoustic data. The start and end of the frequency band is entered into F1 and F2 as shown below:

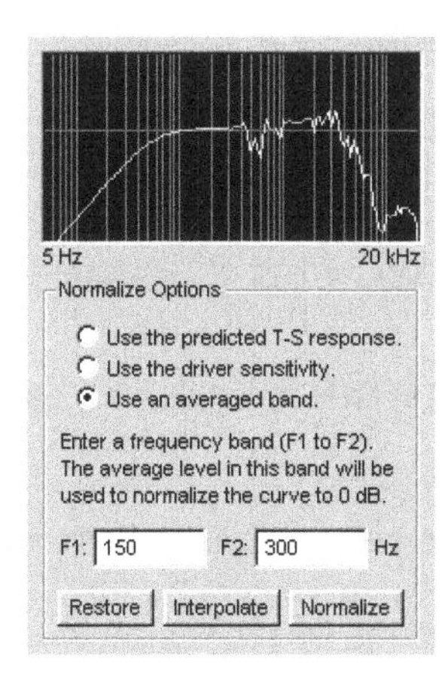

In the above example, F1 = 150 Hz and F2 = 300 Hz so the acoustic data from 150 to 300 Hz will be averaged and this level will be subtracted from each acoustic data point when the "Normalize" button is clicked.

Use a single level (not shown) The final option for normalizing the acoustic data is not listed in the "Normalize Options" box. Like the second and third options, it adjusts the overall level of the acoustic data without changing the shape of the response. It is executed by double-clicking (🖱🖱) on any of the "Level" input boxes. When you do this, a message box will appear to confirm that you want to normalize the acoustic data by subtracting the selected level from each acoustic data point.

Restore The last three acoustic response curves are remembered so that you can "undo" changes when you attempt to normalize the acoustic data. Click the "Restore" button to return to a previous response curve. Clicking the "Restore" button again after the oldest curve has been restored will return you to the most recently stored curve. An acoustic response curve is stored into this "undo" buffer when either the "Interpolate" or "Normalize" button is clicked. The "undo" buffer is not affected by changes to individual acoustic data points with the "Level" input boxes.

Interpolate Clicking the "Interpolate" button will cause X•over Pro to search the acoustic data for all zero values. It will then replace the zero values with estimated values by calculating the slope of surrounding non-zero data points.

2

Normalize Click the "Normalize" button to execute the selected normalization option. These options are described on the previous pages and usually cause the level of the acoustic data to be adjusted to either a predicted response or to the zero line. *Note: The "Normalize" button has one special feature when the "Use the predicted T-S response" option is selected. This feature is activated by holding down the* ⇧Shift *key when the "Normalize" button is clicked to cause the predicted response to be added to the acoustic data rather than subtracted from it as is normal.*

Omit Driver Response in Graphs Check this checkbox to cause the Normalized Amplitude Response graph to ignore the driver acoustic data. This can also be controlled with a checkbox in the Graph Properties window and directly from the graph window by right-clicking (🖱) on the graph and turning off "Include > Driver Acoustic Response" in the popup menu.

Import Acoustic Data Frequency-domain data from several popular measurement systems (including Brüel & Kjaer, CLIO, IMP, LMS, MLSSA, OmniMic, Sample Champion, Smaart, TEF®-20 and TrueRTA) can be imported with the "Import Acoustic Data" button. Clicking it will open the Import Acoustic Response File window so you can select the data file. Several of the file types are identified by the file name extension. Use the "Files of type" drop-down list to select the file type as shown below:

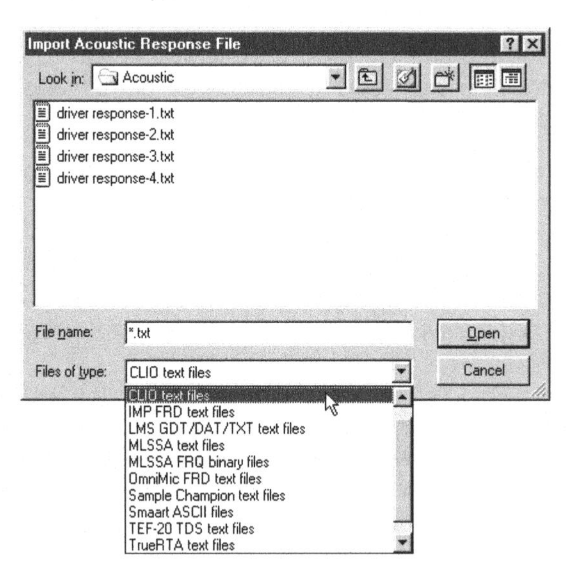

Note: The above Import Acoustic Response File window will vary in appearance based on the version of Windows. For example, the "Files of type" list may not be labeled and may be located to the right of the "File name" input box in Windows 8, 7 and Vista. It is important that the file type be selected with this list so it's type will be correctly identified.

The program will open a Select Data File Type window to prompt the user to select the file type if a file was chosen which the user forgot to identify by the "Files of type" list and X•over Pro cannot identify it by its file name extension.

The acoustic data import filter works in the following manner:

- The import filter will interpolate between data points if some are missing.
- The import filter will average data points if there are extra ones.
- The import filter will give you the option of extrapolating the beginning or end of the data if it begins at a frequency above 5 Hz or ends below 20 kHz. The slope between the first or last five data points will be used if you choose to extrapolate.
- The import filter for some file types will automatically adjust the overall level of the data so that it will be visible in the preview graph.

Note: In most cases, the data will need to be normalized after it has been imported. Please read the preceding instructions about normalization.

Caution: Most acoustic measurements do not include accurate data at very low frequencies. For example, many measurements are grossly inaccurate below 100 Hz. The lowest usable frequency of the measurement will depend on the measurement's frequency resolution, "time window" and the location of nearby reflecting surfaces. These problems can sometimes be overcome with special measurement techniques like the "near field" and "ground plane" techniques. If the low-frequency portion of your acoustic data is not accurate, you should enter zeros for that portion of the acoustic data after the rest of the data has been normalized.

Clear Use this button to set all "Level" settings to zero dB.

Note: The acoustic response of a driver does not have to be entered in order to complete a crossover network or filter design. Most manufactures do not provide it, leaving it up to you, the designer, to measure it yourself.

2

Impedance EQ (Equalization)

The "Impedance EQ" tab contains the settings for a driver's impedance equalization network including, if appropriate, the box parameters. The reason why the box parameters are located on the "Impedance EQ" tab is because their primary purpose is to assist with the design of an accurate impedance equalization network. (Remember, X•over Pro is focused on crossover network design rather than box design.) To learn more about impedance equalization networks and why they are important, see Chapter 3 of the *Crossover Network Designer's Guide*.

The box parameters can be entered even when an impedance equalization network is not desired. This is accomplished by temporarily turning on all of the impedance equalization "Include" options, entering the box information, and then turning off the impedance equalization "Include" options. When box information is present, its effects are always displayed in the graphs.

Notice in the example below that the "Impedance EQ" tab will assist whenever required information is missing. In this case, the "missing data" message identifies two missing parameters (Le and Vb) which must be entered before an impedance equalization network can be designed to account for the resonance peak(s) as influenced by the box.

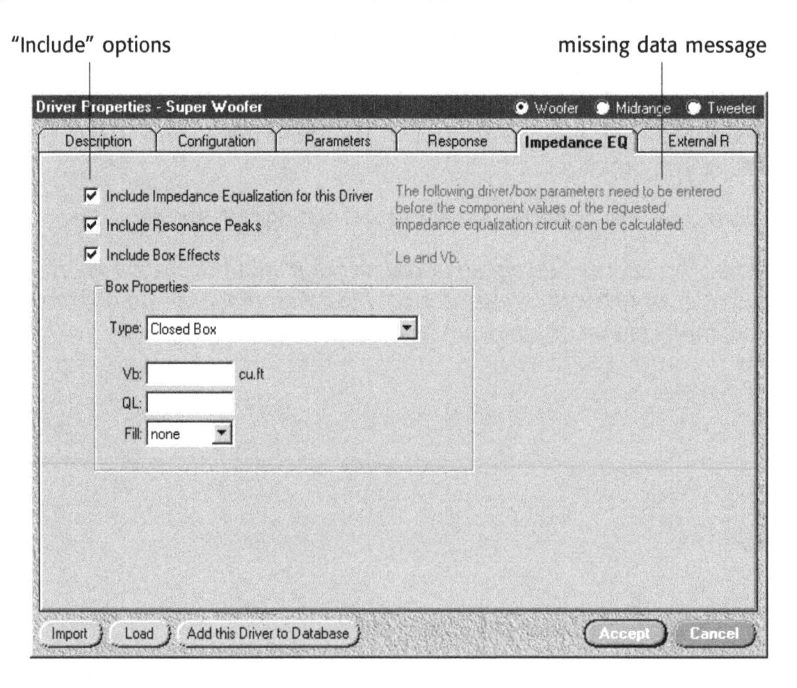

"Include" options Three checkboxes are provided in the upper left corner of the tab. They determine the type of impedance equalization circuit that will be used.

Include Impedance Equalization for this Driver This checkbox enables or disables all impedance equalization for the selected driver. All other impedance equalization options are disabled when this option is turned off. When it is turned on and there is not enough information to calculate the requested impedance equalization network, a red "missing data" message will appear to list the parameters that need to be entered.

Include Resonance Peaks When turned on, this checkbox causes circuitry to be added to the impedance equalization network that flattens the resonance peak(s). It also enables the "Include Box Effects" option. When turned off, the impedance equalization circuitry will equalize only the inductive reactance rise of the driver.

Include Box Effects When turned on, this checkbox causes the impedance equalization network to include the effects of the box on the system resonance. If a vented box is specified, it adds additional circuitry to flatten two resonance peaks. When turned off, the impedance equalization circuitry will assume that only one resonance peak exists and that it is located at the free air resonance of the driver.

X•over Pro provides four different impedance equalization circuits. Each one is shown below along with the list of required driver and box parameters.

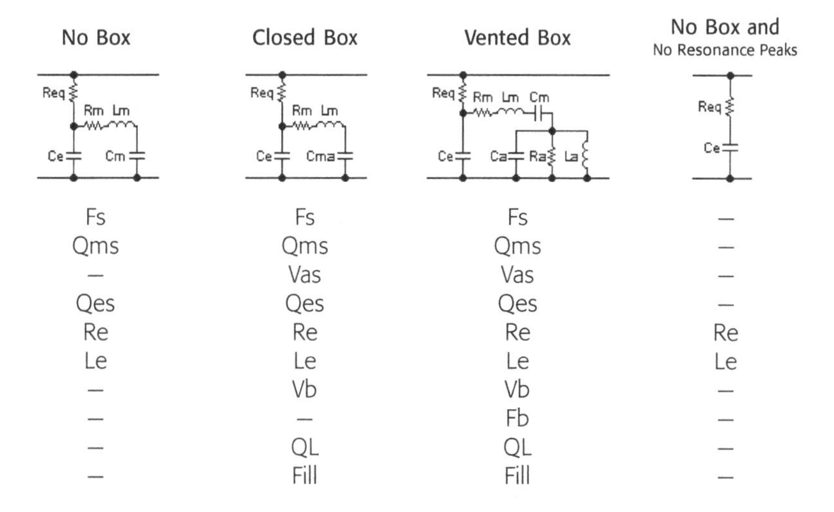

No Box	Closed Box	Vented Box	No Box and No Resonance Peaks
Fs	Fs	Fs	—
Qms	Qms	Qms	—
—	Vas	Vas	—
Qes	Qes	Qes	—
Re	Re	Re	Re
Le	Le	Le	Le
—	Vb	Vb	—
—	—	Fb	—
—	QL	QL	—
—	Fill	Fill	—

The impedance equalization circuit will be included in the circuit diagram of the main window according to the "Include" option settings.

2

Calculate EQ Network This button appears whenever sufficient information is available to calculate or recalculate an impedance equalization network (see the illustration below). If enough information is not available, a red "missing data" message will appear and list the missing parameters. When calculated, the component values of the impedance equalization network will be displayed in the "EQ" tab of the component list in the main window.

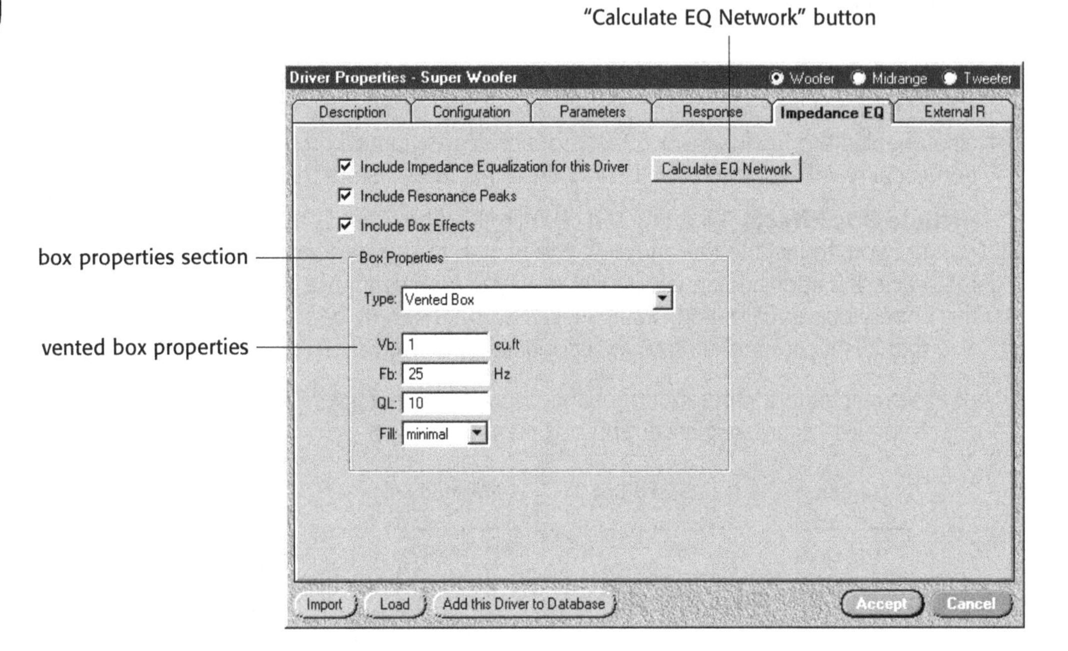

Important: The component values of an impedance equalization network will <u>not</u> change unless the "Calculate EQ Network" button is clicked. This is true even if the driver or box parameters are changed. In this way, you can make manual changes to the impedance equalization network component values on the "EQ" tab of the main window and those changes will be protected when other aspects of the crossover network design are changed. Remember to click the "Calculate EQ Network" button whenever you want X•over Pro to recalculate the impedance equalization network of the selected driver.

Box Properties A "Box Properties" section will appear when the "Include Box Effects" option is turned on. Its contents will vary to reflect the box type that is chosen with the "Type" drop-down list. The illustration above, shows the box properties of a vented box. The illustration on page 120 shows the box properties of a closed box. The box properties of the remaining box types are shown on the next page.

single-tuned bandpass box properties

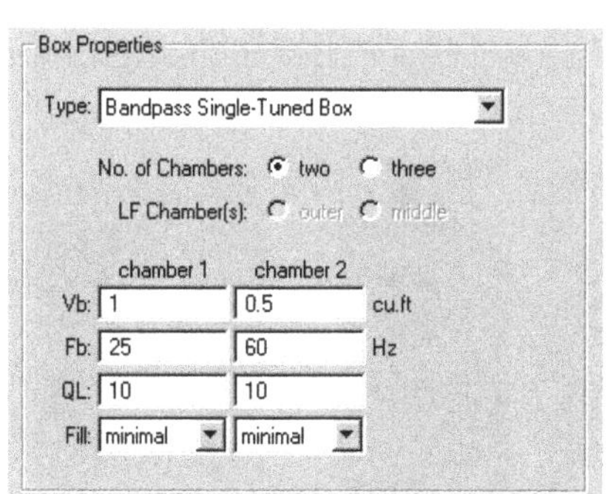

double-tuned bandpass box properties
with triple chambers

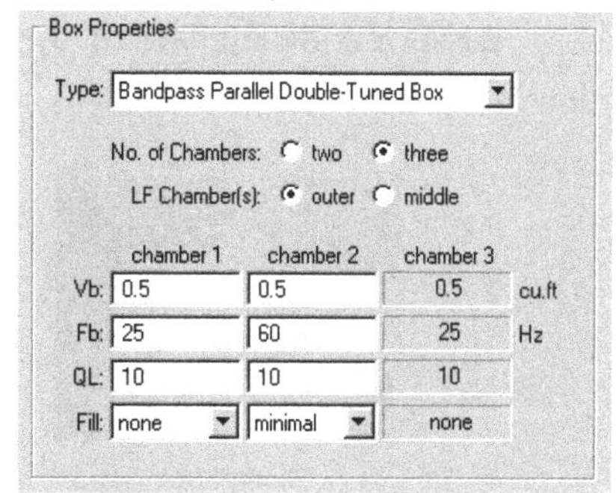

2

passive radiator box properties

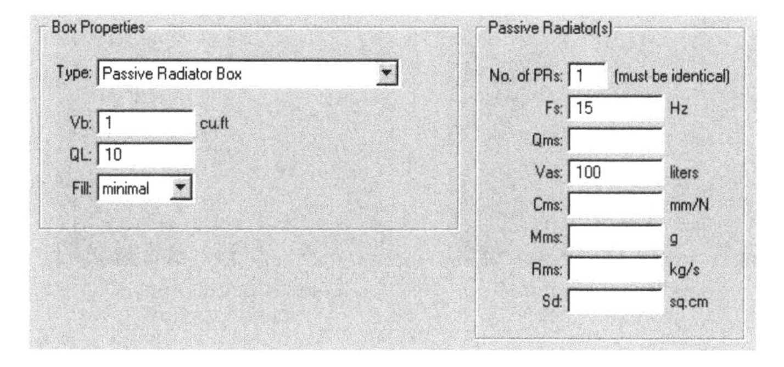

A description of each box parameter follows:

Type A drop-down list which selects the box type. The choices are:

- closed box
- vented box
- vented box with active high-pass equalization filter
- single-tuned bandpass box
- parallel double-tuned bandpass box
- series double-tuned bandpass box
- passive radiator box

A picture of each box type is displayed on the next page.

The parameters in the "Box Properties" section are configured according to the box "Type" selection. The "vented box with active HP EQ filter" is provided for compatibility with

BassBox Pro and BassBox Lite. The active filter settings do not appear and cannot be changed in X•over Pro. However, a box design of this type will be correctly imported from BassBox and the effects of the active filter will be included in the Normalized Amplitude Response graph.

closed box

vented box and vented box with an active HP EQ filter

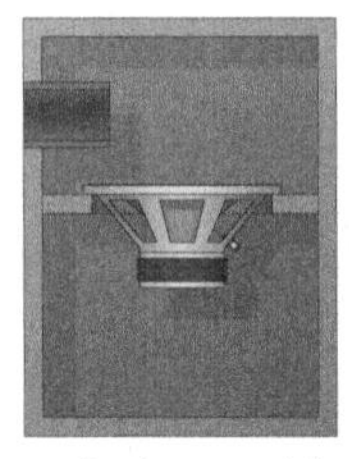

single-tuned bandpass box

parallel double-tuned bandpass box

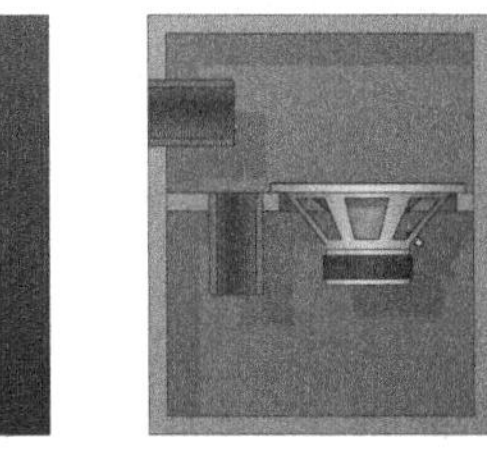

series double-tuned bandpass box

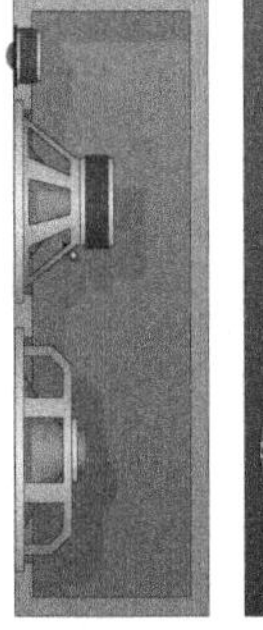

passive radiator box

Important: The impedance equalization networks provided in X•over Pro are optimized for two types of boxes: closed boxes and vented boxes. If another box type is selected, such as a bandpass or passive radiator box, X•over Pro will use the circuit for the vented box. This should provide a reasonably good starting place. You can then plot the impedance response and experiment with the component values until a flat impedance response is achieved.

The box properties are immediately incorporated into the response of the design. This will be visible in the mini preview graph which automatically updates whenever a change is made. This enables X•over Pro to model the box response in addition to the crossover network response. And it gives X•over Pro a basic box design capability. However, a full-featured box design program like BassBox Pro or BassBox Lite is recommended for box design.

No. of Chambers (bandpass boxes only) The total number of chambers in the box. A bandpass box with a single driver can have only two chambers. It is possible to have three chambers when more than one of the same driver is specified in the "Configuration" tab of the Driver Properties window. When three chambers are possible, this option will be activated.

double chamber

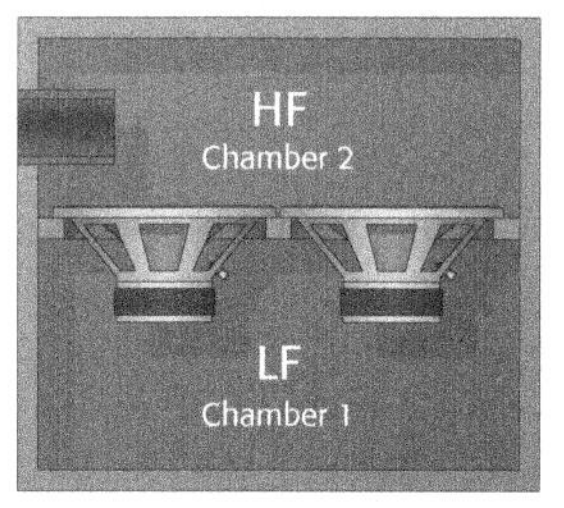

triple chamber

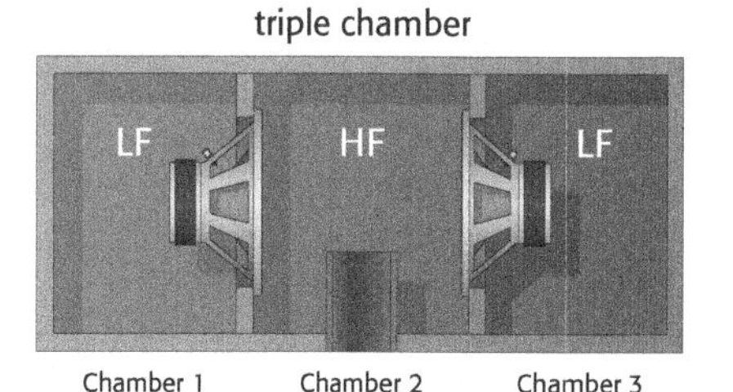

Chamber 1 Chamber 2 Chamber 3

LF Chamber(s) (bandpass boxes only) Identifies which chamber(s) will control the lower frequency (LF) limit of the design. This option is only available when the "No. of Chambers" option above is set to three. With a triple-chamber bandpass box, the outer two chambers are normally used to control the lower frequency limit of the design. The middle chamber is used to control the upper frequency limit. The LF Chambers option allows you to reverse this, causing the outer chambers to control the upper frequency limit and the middle chamber to control the lower frequency limit. *Note: With two-chamber bandpass boxes, Chamber 1 is always considered the low-frequency chamber and Chamber 2 is always considered the upper-frequency chamber.*

LF Chambers = outer

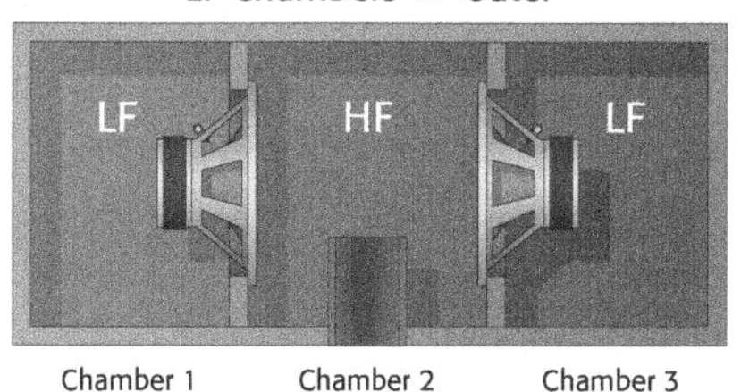

Chamber 1 Chamber 2 Chamber 3

LF Chamber = middle

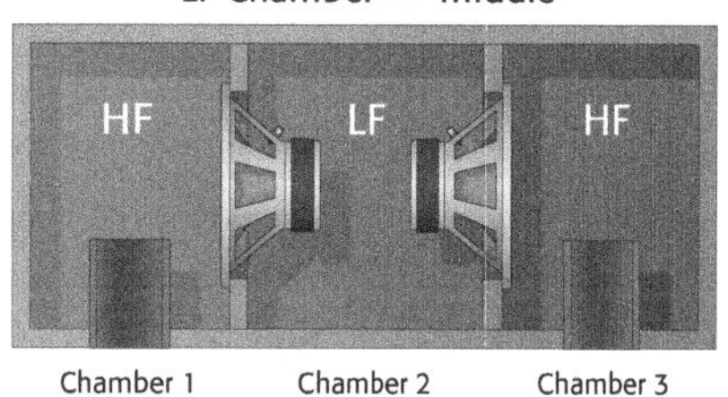

Chamber 1 Chamber 2 Chamber 3

Vb The net internal volume of the box or chamber. This is the volume that remains after subtracting the volume inside the box that is displaced by internal objects like the driver, vent and bracing. Units: liters, cu.cm, cu.m, cu.ft or cu.in.

Fb (vented and bandpass boxes only) The box or chamber resonance frequency. Fb is a product of the net box or chamber volume (Vb) and the vent dimensions. Units: Hz.

Note: The Fb of a sealed low-frequency chamber in a single-tuned bandpass box is the system resonance of the driver and chamber since a vent is not present.

QL The Q of the box leakage losses. The lower the value, the more the losses and visa versa. A high number like 1000 indicates very little leakage loss. When QL has not been entered, it is estimated internally depending on the box volume (Vb). This is done by interpolating between the default small and large box settings in the "Box" tab of the Preferences window. QL can be as low as 5 or less for a very large "lossy" box or as high as 20 for a small well-constructed "tight" box. A QL of 7 is considered to be typical for a vented box of modest size and average quality.

Fill This drop-down list contains the damping setting for the box. The "damping" is the amount of acoustic absorption or "fill" that is placed inside the box (shown below). Five settings are available for the damping parameter: none, minimal, typical, heavy and ignore.

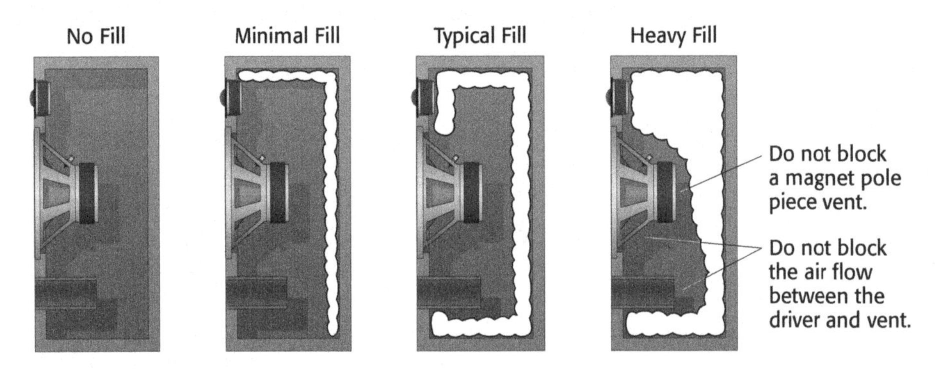

None No absorptive material is added to the interior of the box.

Minimal A modest layer of absorptive material is placed on just one of each pair of opposing walls.

Typical All walls are covered with 1 to 1½ inches (25–40 mm) of absorptive material.

Heavy The box is stuffed full of absorptive material. Care must be taken not to restrict the operation of the driver or, if a vent or passive radiator is present, the air flow between the driver and the vent or passive radiator.

Ignore X•over Pro will ignore the effects of damping material in the box. This means it will use "classical" box calculations and will ignore the effects of Qa and γ (gamma).

Two internal parameters, Qa and γ (gamma), are adjusted whenever a damping selection is made. Qa is the Q of the absorption losses and γ is the ratio of heat at constant pressure to that at constant temperature for the air inside the box with the specified amount of fill. The values used for each fill setting can be adjusted in the "Box" tab of the Preferences window.

Passive Radiator(s)

The "Passive Radiator(s)" section and passive radiator parameters are only available when a passive radiator box "Type" is selected. They are very similar to the driver parameters and only two parameters are needed for the impedance equalization network: Fs and Vas. However, more are required if you want X•over Pro to include the passive radiator box response in the graphs. Here are some of the combinations that can satisfy the latter requirement:

Fs, Qms, Vas, Sd	Fs, Qms, Vas, Cms
Fs, Qms, Cms, Sd	Cms, Mms, Rms, Sd
Fs, Qms, Mms, Sd	

No. of PRs The number of passive radiators. X•over Pro assumes that all of the passive radiators are identical. When more than one passive radiator is specified, X•over Pro will internally calculate and use net values which represent all of the passive radiators. For this reason it is important when specifying more than one passive radiator to enter the following parameters for only one passive radiator.

Fs The free air resonance frequency of the passive radiator. Units: Hz.

Qms The mechanical resonance magnification of the passive radiator at Fs.

Vas The volume of air equal to the mechanical compliance of the passive radiator's suspension. Units: liters, cu.cm, cu.m, cu.ft or cu.in.

Cms The mechanical compliance of the passive radiator's suspension. Units: µm/N, mm/N, cm/N, m/N or in/lb.

Mms The moving mass of the diaphragm or piston including the air load. Units: g, kg or oz.

Rms The mechanical resistance of the passive radiator resulting from its suspension losses. Units: kg/s, lb/s or mohms. (mohms = mechanical ohms)

Sd The diaphragm or piston area of the passive radiator. Units: sq.mm, sq.cm, sq.m or sq.in.

2

External R (Resistance)

The "External R" tab contains the external resistance parameters. Three of the parameters (Rx1, Rg and DF) apply to the system and do not change with the drivers. One parameter (Rx2) applies to a single filter output in the crossover network so it can change with the driver selection. External resistance can affect the function of a crossover network. Sources of it include the amplifier output, speaker cables, connectors and the drivers. *Note: The resistance of each driver (Re) is not included in the external resistance and driver impedance variations are handled with impedance equalization as described previously.*

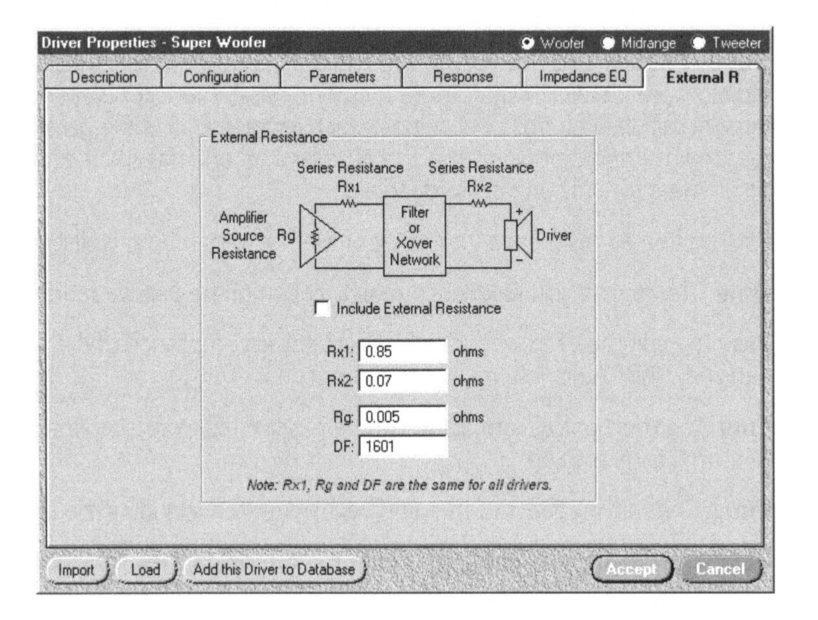

Include External Resistance This checkbox turns on and off the ability of X•over Pro to display the effects of external resistance. When turned on, a "net" value for Qes, BL and Qts will be displayed in the "Parameters" tab to show how they are affected. The "Include External Resistance" checkbox is disabled when Rx1, Rx2 and Rg are all zero.

Rx1 The series resistance of the speaker cables and terminals between the amplifier output and the crossover network input. It does not include the resistance of the amplifier or the crossover network. Since two conductors are required for an audio signal, Rx1 is the total resistance of both conductors. For example, if a 10 foot (3 meter) speaker cable is used between the amplifier and the crossover network, Rx1 would be the resistance of 20 feet (6 meters) of one conductor. There is only one value of Rx1 for the entire network.

Rx2 The series resistance of the speaker cables and terminals between the output of a filter in the crossover network and the input to its corresponding load (the driver). It does not include the resistance of the filter or the load. Like Rx1, it includes the total resistance of both conductors in the speaker cable. There is a separate value of Rx2 for each driver (woofer, midrange driver and tweeter).

Rg The source resistance of the amplifier. This is usually very low with most modern amplifiers. If the driver's impedance (Z) is known, DF is calculated when Rg is entered.

DF The damping factor of the amplifier. If the driver's impedance (Z) is known, Rg is calculated when DF is entered.

2

Polarity

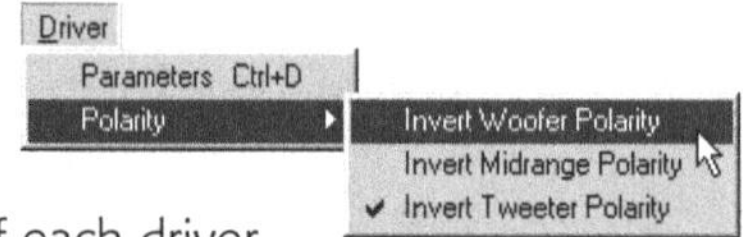

A driver is wired with a normal or positive (+) polarity when a positive audio signal causes its diaphragm to move outward. Reversing the wiring inverts the polarity and causes the diaphragm to move inward. The polarity of each driver is controlled from the Driver menu with the "Polarity" commands shown above.

When should the polarity be inverted? Answer: When the sound waves of two adjacent drivers are close to 180° out of phase at their crossover frequency. Let's look at an example. The graphs below show the amplitude response (left graph) and phase response (right graph) of a 2-way speaker with a 2nd-order 2-way crossover network. Notice the notch in the net response at the crossover frequency (3.154 kHz).

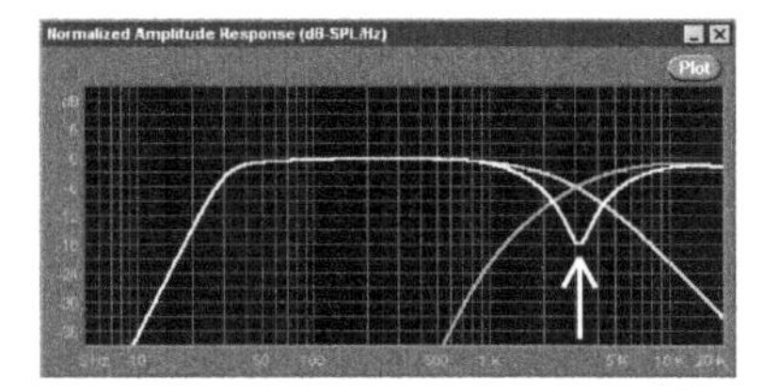

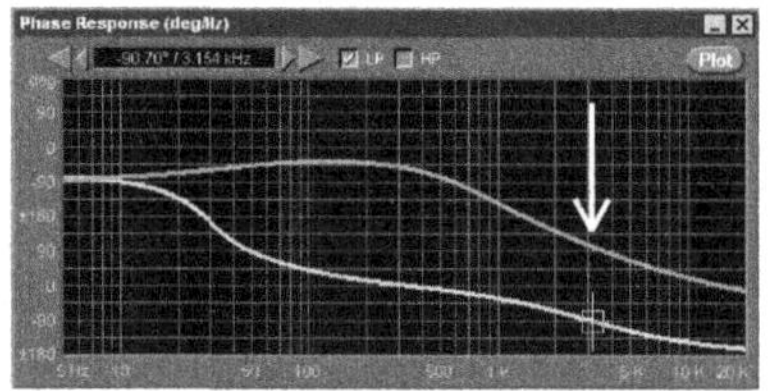

The cursor reveals that the phase of the woofer is –90.70° at the crossover frequency while the tweeter is +103.54°. This is a difference of 194.24° and means that the diaphragms of the woofer and tweeter move in nearly opposite directions at the crossover frequency. Since their sound waves are the same level here, they cancel, creating the notch in the response. To remove the notch, invert the tweeter's polarity as shown in the next two graphs.

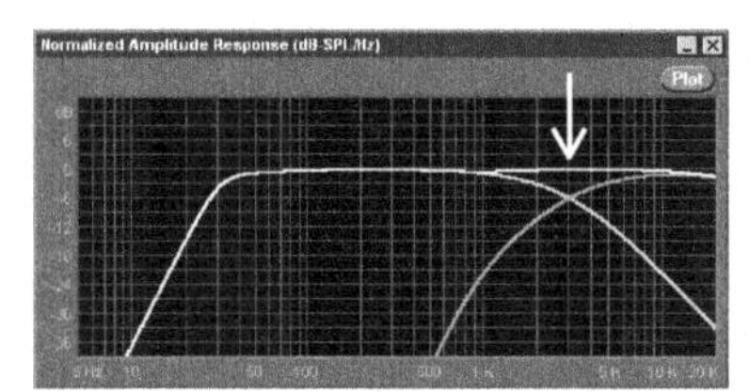

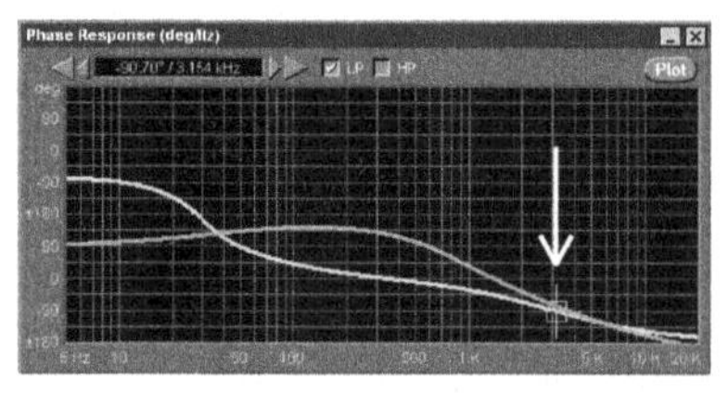

Now the phase response of the drivers are nearly the same at the crossover frequency (woofer: –90.70°; tweeter: –76.46°) and their sound waves sum to a nearly flat response. Remember that X•over Pro shows the system response of the speaker and includes the effects of the drivers and box. The reactive nature of the drivers and box cause the phase of their signals to shift in surprising ways. Phase cancellations will sometimes occur when you do not expect them and visa versa. X•over Pro's graphs can be very helpful in identifying these situations. However, the program must make important assumptions about the driver mounting. Please see the explanation of the net plot line in Chapter 5 (pages 156-158) for more information.

3 Crossover Network / Filter Properties

After the driver data has been entered, relevant impedance equalization networks have been designed and any external resistance has been entered it is time to select a crossover network or filter. Most of the crossover network / filter properties are located on the main window or in its Network menu. Let's begin with the menu and the network type.

Network Type

One of the most important crossover network properties is the Network Type. Three options are provided: "2-way Crossover", "3-way Crossover" or "Separate Filters". They are the first commands listed in the Network menu as shown below (keyboard shortcuts [F2], [F3] or [F4]). Please review Chapter 1 of the *Crossover Network Designer's Guide* earlier in this manual if you are unfamiliar with two-way and three-way crossover networks or filters.

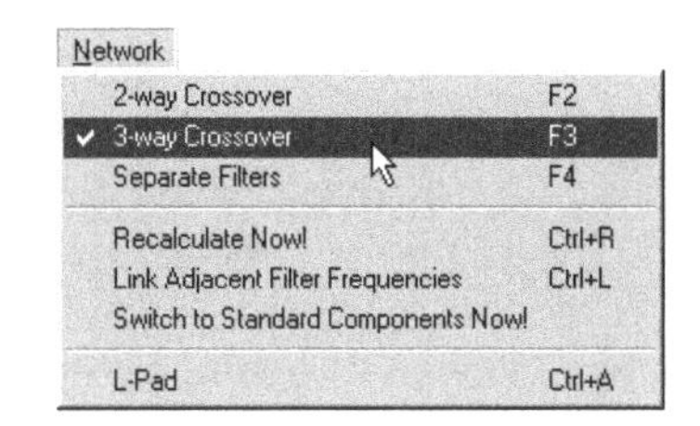

The Network Type you select will depend on your design goals. Sometimes it may be necessary to divide an elaborate or complex crossover network into more than one design. This "divide and conquer" approach can help you design crossover networks for four-way or five-way speakers. Here are a few suggestions:

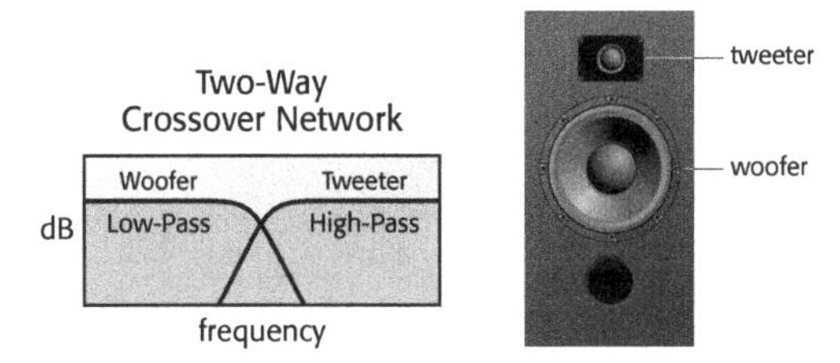

2-way Crossover This Network Type provides a low-pass and a high-pass filter. The low-pass filter is assumed to be connected to an open back driver. The program will assign the "woofer" to this filter; however, you can use any open back driver as long as you enter its parameters into the "woofer" properties of the Driver Properties window (see Chapter 2). For example, you can enter the driver parameters of a small open back mid-bass driver for a two-way "satellite" speaker into the "woofer" properties in the Driver Properties window.

The high-pass filter is assumed to be connected to a sealed back driver and the program will assign the "tweeter" to this filter. *Note: The "midrange" driver properties of the Driver Properties window will be ignored when the "2-way Crossover" is selected.*

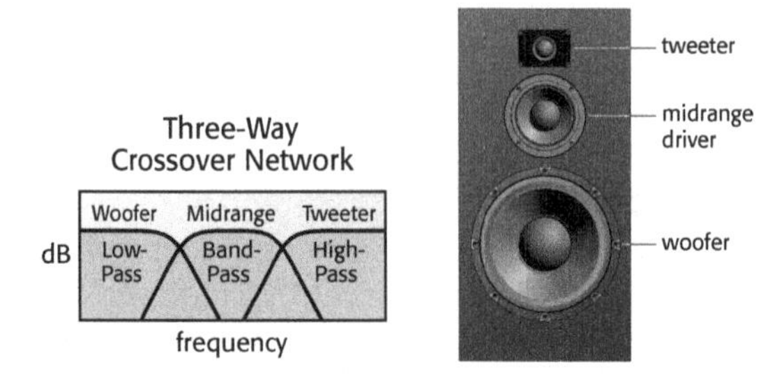

3-way Crossover This Network Type provides a low-pass, band-pass and high-pass filter. The low-pass filter is assumed to be connected to the "woofer". The band-pass filter is assumed to be connected to the "midrange" driver and it can have either an open or sealed back. The high-pass filter is assumed to be connected to the "tweeter".

If you want to design a crossover network for a four-way speaker you will need two band-pass filters. This can be accomplished with two separate "3-way Crossover" network designs. In one design you would ignore the extra low-pass filter and in the second design you would ignore the extra high-pass filter. For example, the first design could include the tweeter/HP filter and the upper midrange driver/BP filter. The second design could include the lower midrange/BP filter and the woofer/LP filter. The response of all four used filters could be plotted one-design-at-a-time using the overlay feature of the graphs. You could also store each design's plot into a graph memory. See Chapter 5 for information about the graph memories.

Feel free to be creative and use the "3-way Crossover" in a way that works best for you.

Separate Filters This selection provides a separate low-pass, band-pass and high-pass filter. None of these filters are assumed to be connected together and so the "net" plot lines will not be available in the graphs. When the "Separate Filters" command is selected, a Select Drivers for Separate Filters window (shown below) will appear to prompt you for a driver assignment for each filter.

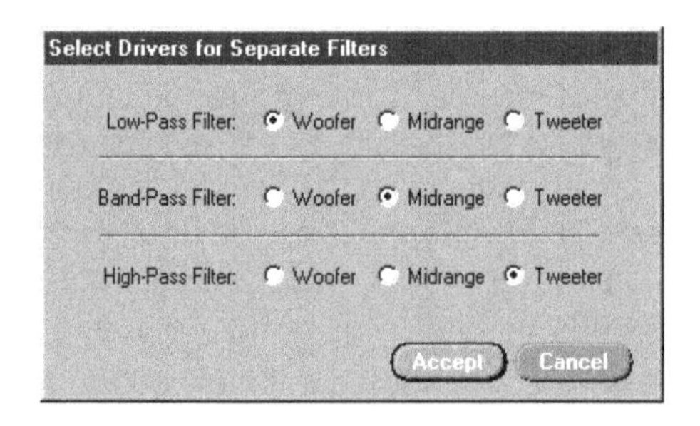

Any time that you want to change a driver assignment, simply choose the "Separate Filters" command again from the Network menu.

It is possible to assign the same driver to all three filters. However, X•over Pro treats each filter as if it is completely separate with its own driver. If, for example, the woofer is selected for both the low-pass and band-pass filters, the program will assume that each filter has its own copy of the woofer. This is readily apparent in the graphs because they plot each filter separately with no net response option.

The main window will be configured appropriately after a network type is selected. The sample below shows a three-way crossover network configuration:

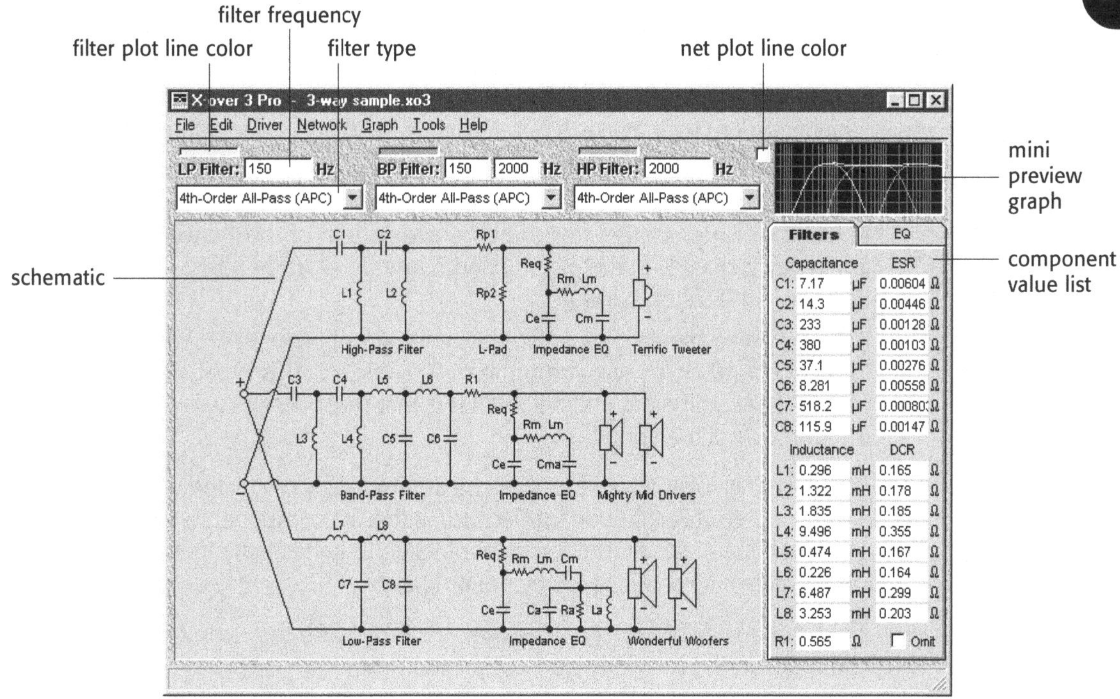

Filter Frequency

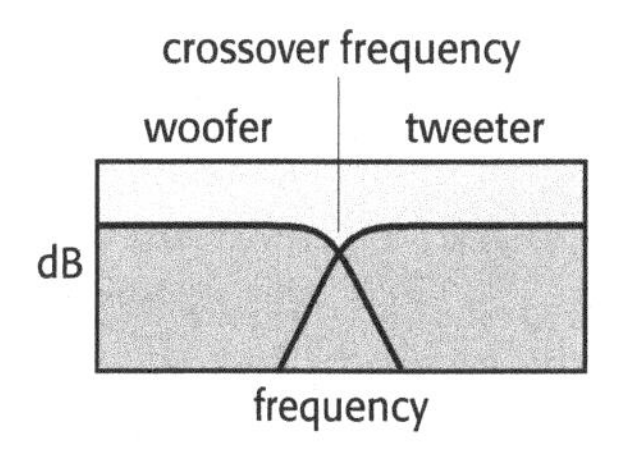

The frequency of each filter is set with the Filter Frequency input boxes at the top of X•over Pro's main window. The Filter Frequency is the cutoff point of the filter and it is the same as the "crossover frequency" for two adjacent filters if both use the <u>same</u> Filter Frequency <u>and</u> Filter Type. If they are the same, the crossover point will usually vary from –3 to –6 dB, depending on the selected Filter Type. (See the "Filter Summary" section in Chapter 1 of the *Crossover Network Designer's Guide* for more information.)

The actual crossover frequency will vary from the Filter Frequency value if dissimilar Filter Frequencies and/or Filter Types are chosen for adjacent filters. The Filter Frequencies may also be inaccurate if the filter component values are manually adjusted on the "Filters" tab of the Component Value List in the main window.

Choosing a Filter Frequency is somewhat of an art. The goal is to help each driver do its best by removing the portion of sound that the driver is unable to efficiently and/or safely produce. A few guidelines are listed below:

3

- Avoid Filter Frequencies that are too close (less than one octave) to the free-air resonance (Fs) of the driver. For example, set the filter frequency no lower than 3000 Hz for a tweeter whose Fs = 1500 Hz. This will also help to protect the driver from too much low-frequency power. This one-octave guideline can sometimes be decreased if a higher order (such as a 4th order) filter is used and it should be increased if a lower order (such as a 1st order) filter is used.
- Select a Filter Frequency that will enable the coverage patterns of two adjacent drivers to blend well. Remember that the coverage pattern of most drivers will narrow as the frequency increases in the piston band of the driver. (See pages 155-156 for information about the piston band.)
- For subwoofers, selecting a Filter Frequency at or below 90 Hz will make it easier to locate the subwoofer away from the main speakers. This is because the wavelengths of sound are so long at and below 90 Hz that most people will not be able to hear the direction of the sound.

New component values for the filters will be automatically calculated whenever a Filter Frequency is changed. This is important because, although component values can be manually edited on the "Filters" tab of the Component Value List, these changes will be lost when the component values are recalculated by the program.

The Filter Frequencies of adjacent filters can be linked with the "Link Adjacent Filter Frequencies" command in the Network menu described next.

Link Adjacent Filter Frequencies

Normally, adjacent Filter Frequencies in a crossover network should be the same. This can be handled automatically with the "Link Adjacent Filter Frequencies" command in the Network menu (keyboard shortcut Ctrl+L). When this feature is turned on, changes made to one Filter Frequency will be automatically mirrored in the adjacent Filter Frequency. *Note: This option is not available when "Separate Filters" is selected.*

Filter Type

The type of each filter is set with the Filter Type drop-down list at the top of X•over Pro's main window. The "type" refers to both the order or slope of the filter and its shape. The available Filter Types change to match the Network Type selection. The choices for two-way and three-way crossover networks are listed below:

Two-way Crossover Networks		Three-way Crossover Networks	
1st-order	Butterworth (APC & CPC)	1st-order	All-Pass (APC)
	Solen Split –6 dB		Constant-Power (CPC)
			Solen Split –6 dB
2nd-order	Bessel	2nd-order	All-Pass (APC)
	Butterworth (CPC)		Constant-Power (CPC)
	Chebychev	3rd-order	All-Pass (APC)
	Linkwitz-Riley (APC)		Constant-Power (CPC)
3rd-order	Butterworth (APC & CPC)	4th-order	All-Pass (APC)
4th-order	Bessel		Constant-Power (CPC)
	Butterworth (CPC)		
	Gaussian		
	Legendre		
	Linear Phase		
	Linkwitz-Riley (APC)		

When "Separate Filters" is selected instead of a two-way or three-way crossover network, a combination of the above Filter Types will be available. The low-pass and high-pass filters will use the two-way choices and the band-pass filter will use the three-way choices. See the "Filter Summary" in Chapter 1 of the *Crossover Network Designer's Guide* earlier in this manual for a description of each filter.

Choosing a Filter Type, like the Filter Frequency, is somewhat of an art. The goals are similar: help each driver do its best by removing the portion of sound that the driver is unable to efficiently and/or safely produce. A few guidelines are listed below:

- Select a Filter Type that will provide adequate isolation between adjacent drivers. 4th-order filters provide the best isolation and 1st-order filters provide the least isolation.
- Select a Filter Type that will complement the speaker design. For example, often a 3rd-order Butterworth or a 4th-order Linkwitz-Riley Filter Type is chosen for a D'Appolito mid-tweeter-mid speaker design.
- Select an APC (all-pass crossover) Filter Type when a flat on-axis amplitude response is desired. This is generally considered the best for most situations.
- Select a CPC (constant-power crossover) Filter Type when a flat power response is desired. Although less common, these types of filters are sometimes beneficial when the speaker will be used off-axis or in a reverberant environment.

New component values for the filters will be automatically calculated whenever a Filter Type is changed. This is important because, although component values can be manually edited on the "Filters" tab of the Component Value List, these changes will be lost when the component values are recalculated by the program.

Filter Plot Line Color

Each filter has a separate color setting so it can be distinguished in the graphs. The Filter Plot Line Color indicators serve both to indicate the plot color of each filter and allow you to change them. By clicking (🖱) on them the color will advance one at a time through a 12-color palette with each single click. The first ten colors of the palette are fixed (red, orange, yellow, greenish-yellow, green, cyan, blue, magenta, white and light grey). But the last two colors are "custom" colors and you can make them any color you'd like. The default plot colors can be selected with the "Graph" tab of the Preferences window (see Chapter 11).

Changing a plot line color setting will not change the color of plot lines that already exist in the graph windows. It will only affect future plotting.

Net Plot Line Color

A "net" plot line is available when a "2-way Crossover" or "3-way Crossover" network is selected. It shows the summed response of all filters in the design. See Chapter 5 (pages 156-158) and Chapter 11 (pages 219-221) for information on how the net plot line is calculated and configured. The "net" plot line is not available for "Separate Filters".

The Net Plot Line Color indicator serves to both indicate the plot color and provide a means of changing the color. Simply click (🖱) on it to advance through the 12-color graph palette like the Filter Plot Line Color indicators. Additional information on the graph color palette and the net plot line is available in Chapter 5.

Recalculate Now!

At some point during the design process it may become necessary to force a recalculation of the component values of the crossover network or filters. You may need to recalculate the crossover network after changing one or more driver parameters or you may want to restore the original values after they have been changed on the "Filters" tab of the Component Value List of the main window. Select the "Recalculate Now!" command (Ctrl+R) in the Network menu shown below to recalculate the component values.

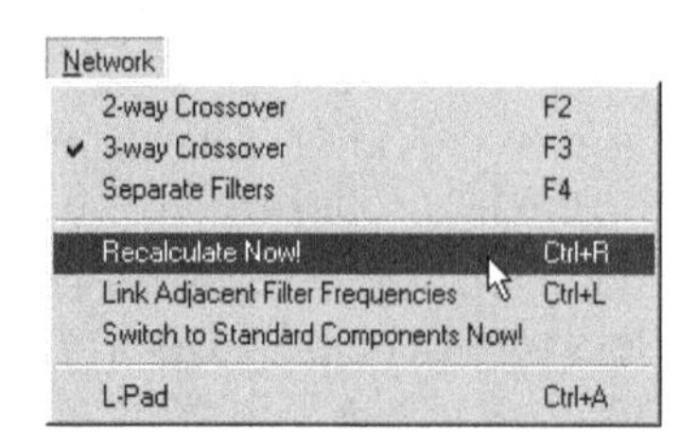

Executing this command will not affect the component values of the impedance equalization networks or L-pads. To recalculate the impedance equalization networks, click the "Calculate EQ Network" button on the "Impedance EQ" tab of the Driver Properties window. You will need to do this for each driver that you want to recalculate. To recalculate the component values of the L-pads, open the L-Pad Attenuator window, re-enter the "Attenuation" level and then click the "Add" button. This will also need to be done for each driver.

Remember that the component values of a crossover network filter will be automatically recalculated whenever its Filter Frequency or Filter Type is changed. The reason the component values are not recalculated at other times is to protect manual changes which you may have made to them with the "Filters" tab of the Component Value List on the main window. Use the "Recalculate Now!" command whenever one filter in a crossover network is changed so that adjacent filters will be refreshed (their values are sometimes influenced by adjacent filters).

Switch to Standard Components Now!

X•over Pro always tries to calculate precise values for components. For example, the first capacitor (C1) in the illustration on page 133 has a value of 7.17 µF. Unfortunately a capacitor with this exact value may not exist. There are two ways to deal with this. One way is to combine two or more capacitors to achieve the desired value. A Parallel-Series Value Calculator is included in the Tools menu to aid with such calculations (see page 205).

The second way of dealing with this problem is to substitute the closest standard-value component and see if the result is satisfactory. The "Switch to Standard Components Now!" command in the Network menu is provided to do just that. It will immediately substitute the closest standard value for all capacitors, inductors and resistors in the crossover network or filters, impedance equalization networks and L-pads.

If the standard values are unsatisfactory, how do you restore the precise values again? Execute the "Recalculate Now!" command in the Network menu to recalculate the crossover network components. Return to the Driver Properties window and use the "Calculate EQ Network" button on the "Impedance EQ" tab to restore each impedance equalization network. Return to the L-pad Attenuator window to restore each L-pad. *Note: You will need to restore the impedance equalization network and L-pad of each driver separately.*

The "standard" values used by X•over Pro are gleaned from electronics component manufacturers and catalogs.

3

Schematic

The most prominent feature of the main window is the large schematic or circuit diagram depicting the crossover network, driver impedance equalization networks, L-pads and driver wiring configurations. It is automatically updated whenever the design is changed. Four display styles are available as shown below:

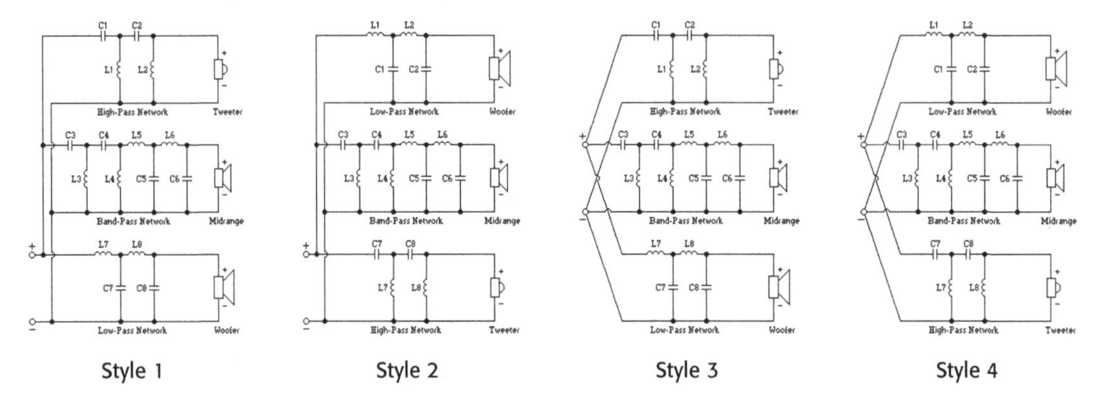

Style 1 Style 2 Style 3 Style 4

The styles differ in the way the input is drawn and whether the tweeter or the woofer is placed on top. Double-click anywhere on the schematic to toggle through each of the styles. The schematic will print the same way it is drawn on screen. The default style is set with the "General" tab of the Preferences window (see Chapter 11).

When multiple drivers are specified, there may be an occasion when the schematic is too wide to fit within its allotted area on the screen. When this happens, you can scroll the schematic left and right to view all of it. This is shown in the illustration at the top of the next page. Notice that the pointer is changed to a left-right arrow when the left mouse button is pressed to show that the schematic is ready to be scrolled. To move the schematic left and right, drag the mouse left and right while the left button is pressed.

Mini Preview Graph

The mini preview graph displays the normalized amplitude response of the system from 5 Hz to 20 kHz with a vertical scale of 9 dB/division. It replots automatically whenever the design is changed. In this way it provides real-time feedback of the network while you work on it. You can force it to replot by left clicking on it. Right clicking on it will display a pop-up menu containing some of the same options of the full-size graphs. Changing any of these options will affect all X•over Pro graphs.

Changing the plot line color of one of the filters or the net plot line will immediately cause the mini preview graph to be refreshed with the new plot color.

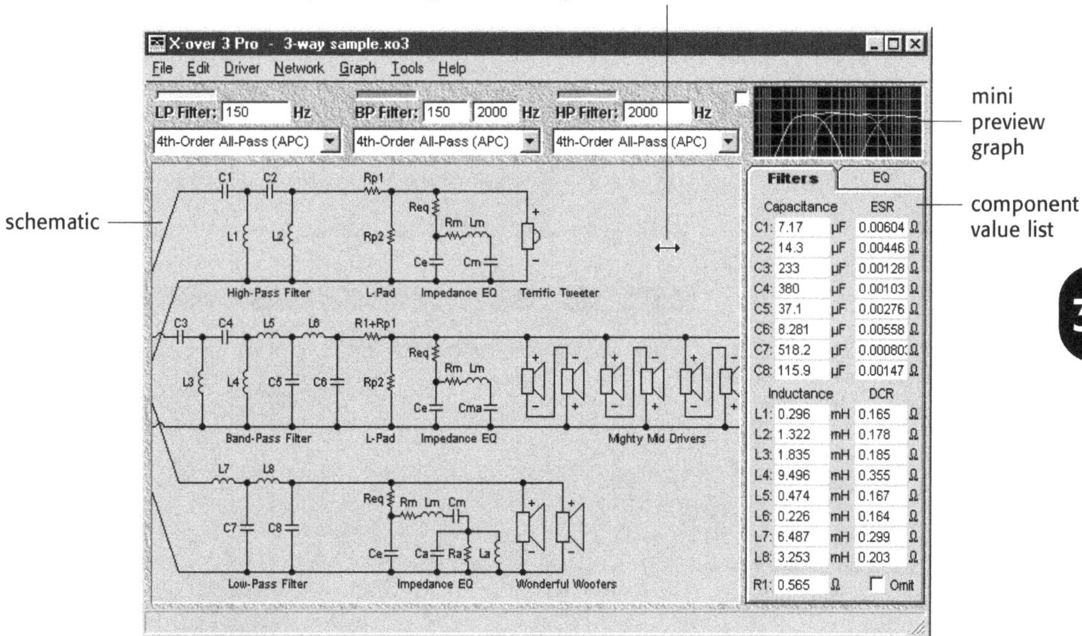

Component Value List

A list of all capacitors, inductors and resistors in the crossover network, impedance equalization networks and L-pads is provided in the Component Value List of the main window. The component information is divided between two tabs (shown at right). The "Filters" tab contains the components for the filters which make up the crossover network. The "EQ" tab contains the components for the driver impedance equalization networks and L-pads.

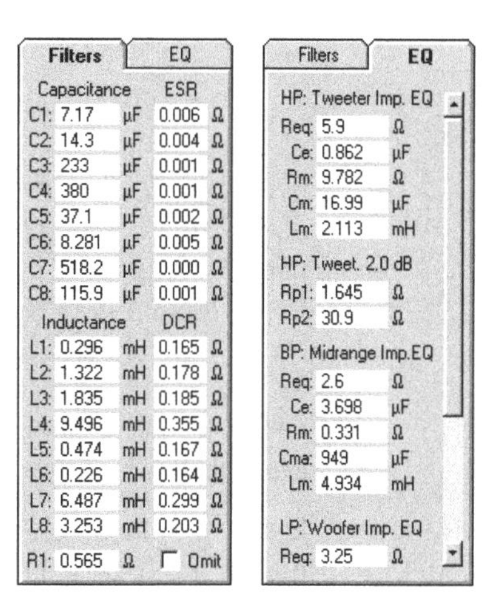

The equivalent series resistance (ESR) and DC resistance (DCR) are required for the capacitors and inductors in the crossover network filters but not for the impedance equalization networks. X•over Pro will automatically estimate the ESR and DCR whenever possible. They can also be manually entered.

In order to estimate the ESR and DCR the program must know what kind of capacitor or inductor is used. This can be set with the Filter Component Resistance Estimator window. To open it, right-click (🖱) on one of the ESR or DCR input fields on the Component Value List or select the "Component Resistance Estimator" command from the Tools menu (keyboard shortcut [Ctrl]+[E]).

Tools
Component Resistance Estimator Ctrl+E
Parallel-Series Value Calculator
Color Value Decoder
Notch Filter Designer
Start BassBox Pro

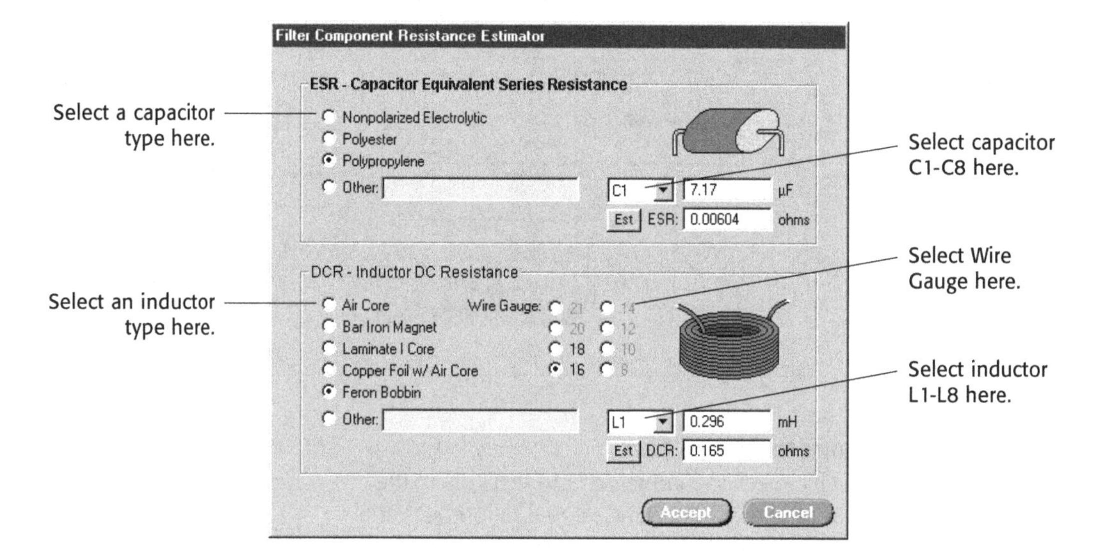

Begin by selecting the desired component (C1-C8 or L1-L8). This is done with the drop-down list to the left of the component value. Next, select the component type. Capacitors offer "Nonpolarized Electrolytic", "Polyester", "Polypropylene" and "Other". Inductors offer "Air Core", "Bar Iron Magnet" core, "Laminate I Core", "Copper Foil w/ Air Core", "Feron Bobbin" and "Other". If the DCR is being estimated you will also need to select the "Wire Gauge" of the inductor. DCR estimation is only available for Wire Gauge values in black (not grey). Finally, click the "Est" (Estimate) button to estimate the resistance. The "Est" button is not available when "Other" or an unsupported inductor Wire Gauge (grey value) is selected.

Remember that the ESR and DCR can always be entered manually in the appropriate input box on the "Filters" tab of the main window. To remove the effects of ESR and/or DCR from the network, enter a very small value like 0.001 ohms.

The default Capacitor Type, Inductor Type and Wire Gauge can be set with the Preferences window (see Chapter 11).

4 L-Pad Attenuator

The L-Pad Attenuator window (shown below) provides a versatile way to design an L-pad attenuator circuit for one or more drivers. To open it, select the "L-Pad" command from the Network menu of the main window or use the keyboard shortcut Ctrl+A.

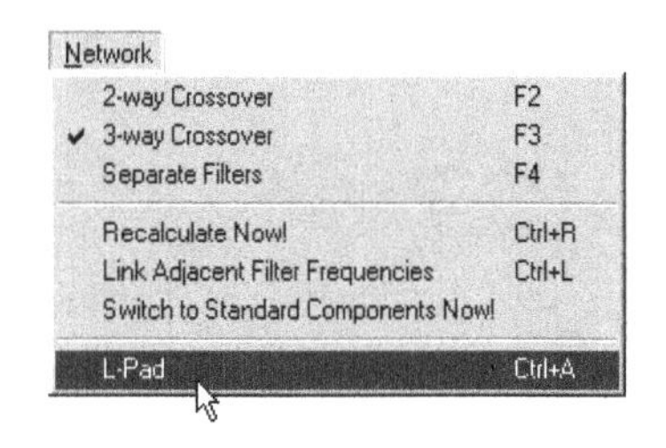

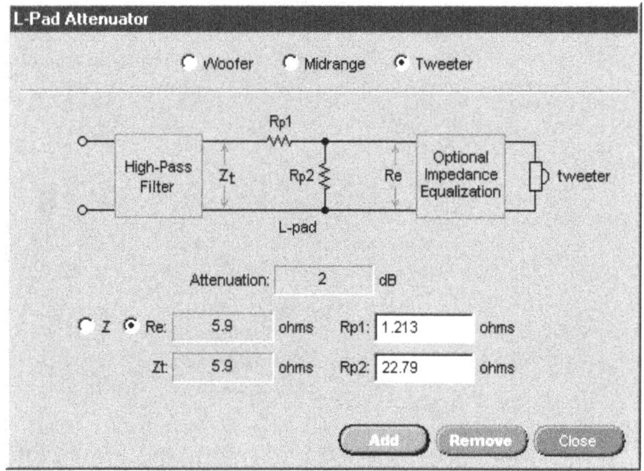

L-pads are used to attenuate loud drivers so that they match the sound level of the quieter drivers in a speaker. This is explained in the Chapter 4 of the *Crossover Network Designer's Guide* earlier in this manual.

An L-pad contains two resistors: Rp1 and Rp2. By default, the L-Pad Attenuator window is configured so that Rp1 and Rp2 can be entered by the user. This is why they have white input boxes as shown above. In this mode the attenuation of the L-pad and the total load impedance (Zt) are automatically calculated whenever Rp1 or Rp2 are changed. Notice also that the user can select whether the driver's nominal impedance (Z) or DC resistance (Re) is used as a basis for these calculations.

Designing an L-pad

Begin by selecting the desired driver at the top of the L-Pad Attenuator window. *Note: Some drivers may not be available if both their impedance (Z) and DC resistance (Re) are unknown. Also, the midrange driver is not available when a "2-way Crossover" network is selected in the Network menu.*

An easy way to design an L-pad is to enter the desired attenuation and let X•over Pro calculate the value of resistors Rp1 and Rp2. To do this, you will need to click in the "Attenuation" input box to reconfigure the L-Pad Attenuator window for manual control of the attenuation level and desired total impedance (Zt). Notice on the next page how the Attenuation and Zt input boxes change to white when they are available for input. Notice also that Rp1 and Rp2 switch to a grey background showing that they can no longer be entered and will instead be calculated.

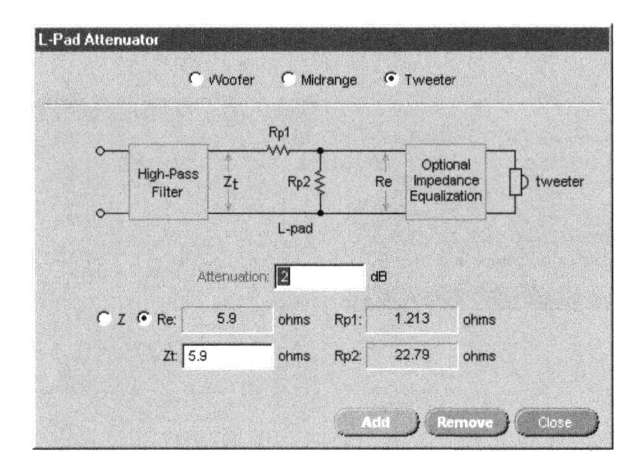

Next, enter the desired attenuation level into the "Attenuation" input box. This should be a positive value because "attenuation" implies a decrease in level. Then select whether you want the driver's nominal impedance (Z) or DC resistance (Re) to be used as the basis for the calculations. The Z or Re value will be displayed. Normally, the same basis should be used for L-pad calculations as for the rest of the network. *Note: Neither the value of Z or the value of Re can be set here. To change either parameter value, use the "Parameters" tab of the Driver Properties window (see page 112).*

Finally, enter the total impedance (Zt) that you want the crossover network filter to "see". Normally, Zt should be set to the same value as the driver's Z or Re (whichever one is used as the basis for the calculations). Whenever an L-pad is present, the Zt value will be used when a low-pass, band-pass or high-pass crossover network filter is calculated. The Rp1 and Rp2 values will be immediately calculated whenever the attenuation level or Zt value are changed.

Important: Whenever an L-pad is added, changed or removed, the response will change. Often the corresponding crossover network filter will need to be recalculated to match the new load impedance (Zt). The crossover network filters are recalculated with the "Recalculate Now!" command in the "Network" menu or with the keyboard shortcut Ctrl+R.

Attenuation The decrease in sound level created by the L-pad in dB (decibels). It must be a positive value because it is the amount of decrease and a "negative decrease" would be the same as an increase and is not allowed. If the Attenuation input box is grey, you will need to click on it first to reconfigure the L-Pad Attenuator window for attenuation, Zt input.

In this mode the Rp1 and Rp2 values will be recalculated whenever the attenuation or Zt values are changed.

Z, Re Use the "Z" and "Re" buttons to select the nominal impedance (Z) or DC resistance (Re) as the basis for the L-pad calculations. Normally, this setting should match the default setting of the program. (The default setting can be controlled with the "General" tab of the Preferences window. See page 211.) *Note: Neither the value of Z or the value of Re can be set here. To change either parameter value, use the "Parameters" tab of the Driver Properties window (see page 112).*

When only one of these values is known, the unknown value will be estimated. X•over Pro assumes that Z is 1.2 times larger than Re. If Z is selected and only Re is known, Z will be estimated by multiplying Re by 1.2. Conversely, if Re is selected and only Z is known, Re will be estimated by dividing 1.2 into Z. If you want to use a different value, you will need to enter it on the "Parameters" tab of the Driver Property window for the corresponding driver.

Zt The total or net load impedance in ohms. This is the load impedance that the corresponding low-pass, band-pass or high-pass crossover network filter will "see". If an impedance equalization network is used (recommended) then the load impedance should remain relatively constant with frequency and crossover filter design will be more predictable. If no impedance equalization network it used, the impedance may vary with frequency and Zt should be considered a nominal value only (like the driver's nominal impedance, Z).

If the Zt input box is grey, you will need to click on it first to reconfigure the L-Pad Attenuator window for Zt, attenuation input. In this mode the Rp1 and Rp2 values will be recalculated whenever Zt or the attenuation is changed.

Note: It is possible to enter a desired value for Zt that cannot be achieved. When this happens, Rp2 will become negative and will be highlighted in red. Enter a different value for Zt until Rp2 is positive. X•over Pro will not allow an L-pad to be used if one or more values are negative.

Rp1 The series resistor of the L-pad in ohms. If the Rp1 input box is grey, you will need to click on it first to reconfigure the L-Pad Attenuator window for Rp1, Rp2 input. In this mode the attenuation level and Zt values will be recalculated whenever Rp1 or Rp2 are changed.

Note: It is possible to enter a value for Rp1 only. In such cases, the level will still be attenuated but without Rp2 the attenuator will not be an "L-pad".

Rp2 The parallel resistor of the L-pad in ohms. If the Rp2 input box is grey, you will need to click on it first to reconfigure the L-Pad Attenuator window for Rp2, Rp1 input. In this mode the attenuation level and Zt values will be recalculated whenever Rp2 or Rp1 are changed.

Note: It is possible to enter a value for Rp2 only. However, as long as Rp1 is not used, there will be no attenuation and only the net load impedance (Zt) will change.

Add Calculating an L-pad is only the first step—immediately after it has been calculated, it will still need to be "added" to the design with the "Add" button. This second step of clicking the "Add" button will cause the L-pad circuit to be added to the schematic in the main window and it will add Rp1 and Rp2 to the list of components on the "EQ" tab of the main window. The "Add" button should also be used after an L-pad is edited to replace the old values of Rp1 and Rp2 with the new ones. The graphs will not show the effects of the new L-pad values until after the "Add" button has been used.

Only one L-pad is allowed per driver and it is located in the circuit between the output of the corresponding crossover network filter and, if present, the input of the impedance equalization network.

Remove If desired, use the "Remove" button to remove an L-pad from the design.

Changing Rp1 and Rp2 on the "EQ" Tab

After an L-pad is added to a design, the attenuation level and the values of Rp1 and Rp2 will be listed along with any impedance equalization component values on the "EQ" tab on the right side of X•over Pro's main window. This is shown below:

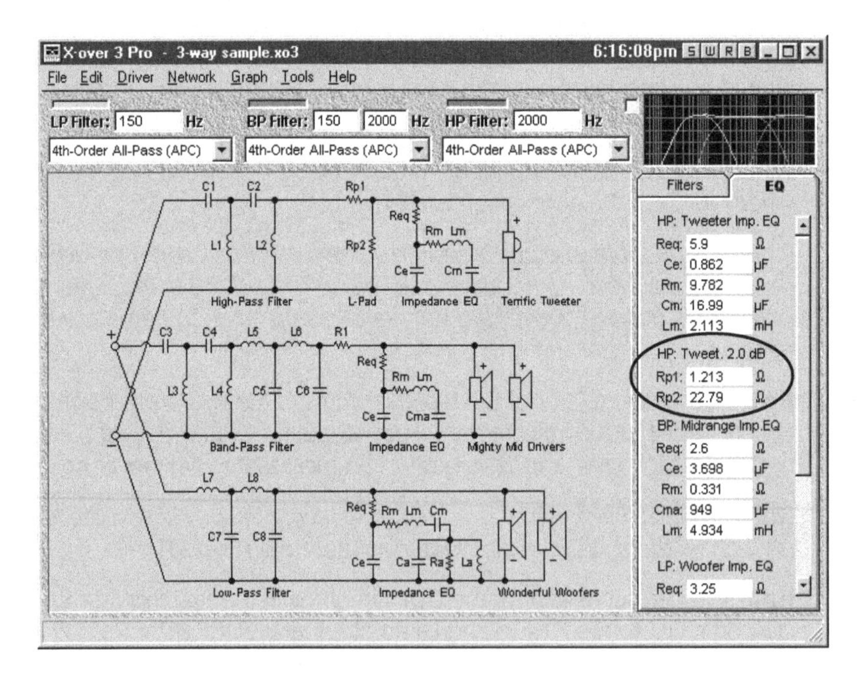

These values can be changed here any time you want to experiment with a different value for Rp1 or Rp2. The attenuation level will be automatically recalculated and displayed whenever Rp1 or Rp2 are changed. This is a powerful tool for tweaking a design—but don't forget that changes to Rp1 and Rp2 may change the impedance "seen" by the filters in the crossover network, adversely affect the response and require the network component values to be recalculated. This will be readily apparent in the mini preview graph if sufficient driver parameters are available.

How much attenuation is needed?

One way to answer this question is to compare the sensitivities of each driver. For example, if a two-way speaker has a woofer with a sensitivity of 90 dB and a tweeter with a sensitivity of 93 dB, then the tweeter will probably need 3 dB of attenuation. However, there is a better way to determine the required attenuation—observe the level of each driver in the Normalized Amplitude Response graph. Why is the graph better? Because the graph also includes the insertion loss of the various crossover network filter components and because the sensitivity ratings of some manufacturers are not accurate.

Two examples are shown below. The first graph shows the response before an L-pad is added. The second graph shows the response after a 2 dB L-pad is added to the tweeter.

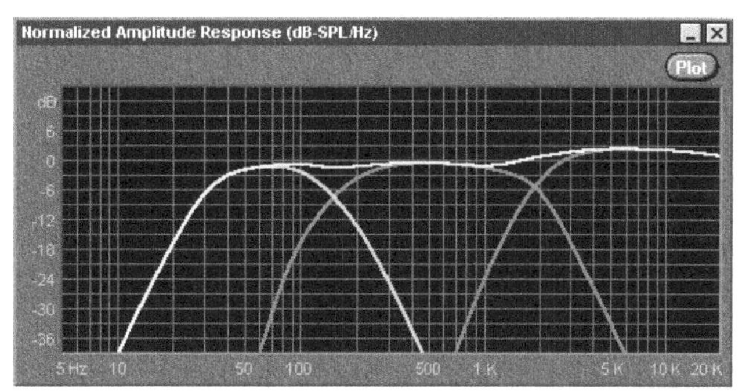

no L-pad

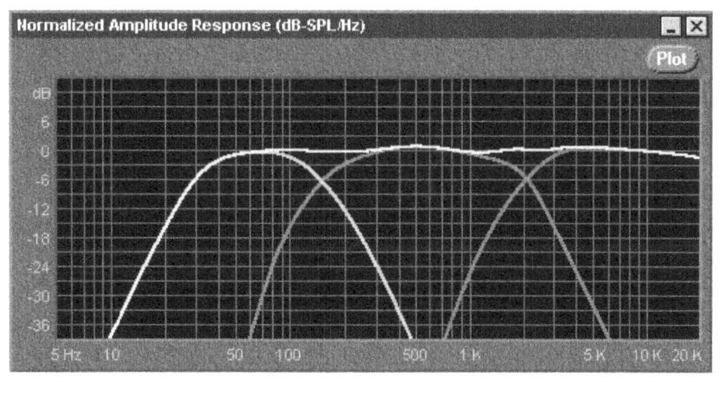

2 dB L-pad added to tweeter

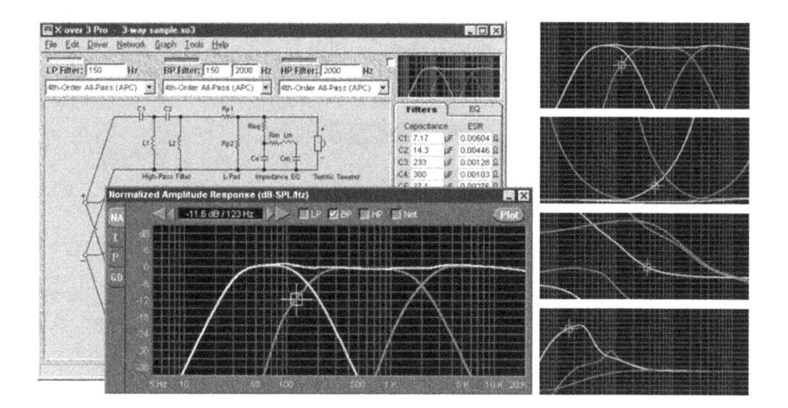

5 Evaluating Performance

A skillful crossover network design is a matter of balancing the various compromises that must be made. No speaker system yet devised is perfect. In fact, speakers are still considered the weakest "links" in the high-fidelity "chain". Always begin the design process by honestly evaluating the intended purpose and use for the speaker. Remember that purpose when evaluating its performance. This will help you decide which are the most important criteria to achieve.

The main window of X•over Pro includes a mini preview graph that depicts the normalized amplitude response of the system from 5 Hz to 20 kHz. Because it automatically updates whenever the design is changed, it helps you immediately evaluate the merits of your design decisions. But this mini preview graph only touches the surface of X•over Pro's powerful graphing capabilities.

This chapter explores the features of X•over Pro's four full-size graphs and their use in evaluating the performance of a crossover network design.

Graph Modes

There are two graph modes: "single window" mode and "individual windows" mode. The mode can be selected from the Graph menu of the main window (shown below). The default graph mode can be set with the "Graph" tab of the Preferences window.

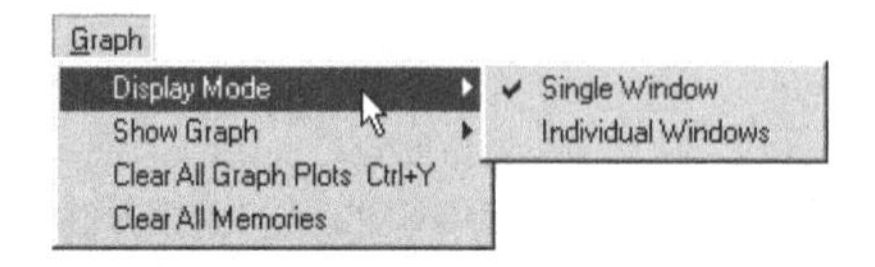

Single Window Mode

The "single window" mode combines all four graphs into a single window. It is recommended for computers with a low graphics resolution (less than XGA or less than 1024 x 768 pixels) because it economizes space on the Windows desktop. It also uses less system resources which may be helpful if your computer doesn't have much memory and/or you often run other large applications at the same time as X•over Pro.

"single window" mode

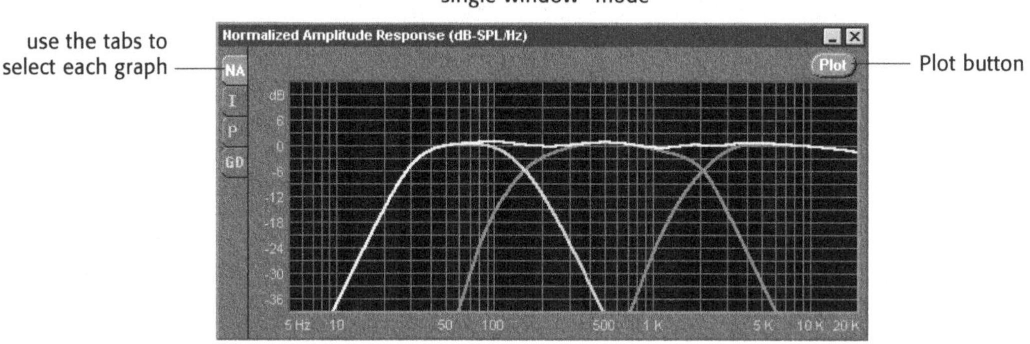

A row of vertical tabs along the left side of the window is used to select a graph. Their abbreviations represent:

NA – normalized amplitude response
I – system impedance
P – phase response
GD – group delay

Individual Windows Mode

The "individual windows" mode displays each graph in its own window, allowing multiple graphs to be viewed simultaneously. This can be a great time-saver when you need to view several things at the same time. However, the graphs can quickly crowd the Windows desktop unless your computer graphics system has a high resolution (like 1600 x 1200 pixels). The "individual windows" mode also provides one other advantage: a large-size option. This allows you to enlarge the size of one or more of the graph windows.

"individual windows" mode

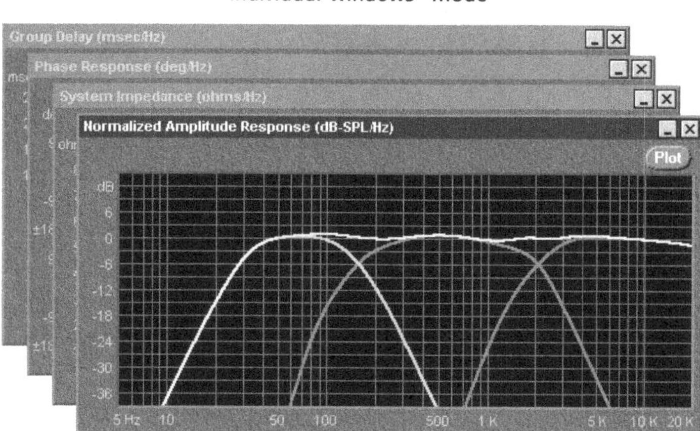

Graph Features

There are many graph features and many of them have default values which can be selected in the Preferences window. The eight options listed below can only be configured with the "Graph" tab of the Preferences window (see pages 219-221):

- the individual filter plot lines can be turned on and off
- include / ignore the phase in the net amplitude response plot line
- the plot line width
- overlay / don't overlay the plot lines
- the grid line darkness
- two custom plot colors
- the offset of the net plot line above or below the individual filter plot lines
- the shape of the graph cursor

The majority of the graph options can be configured from the graph popup menu or the Graph Properties window. These two different methods are provided so you can choose the one that you prefer.

5

5

Graph Popup Menu
The graph popup menu (shown below) is accessed by right-clicking (🖱) on the graph. Most of the graph options can be controlled from this menu.

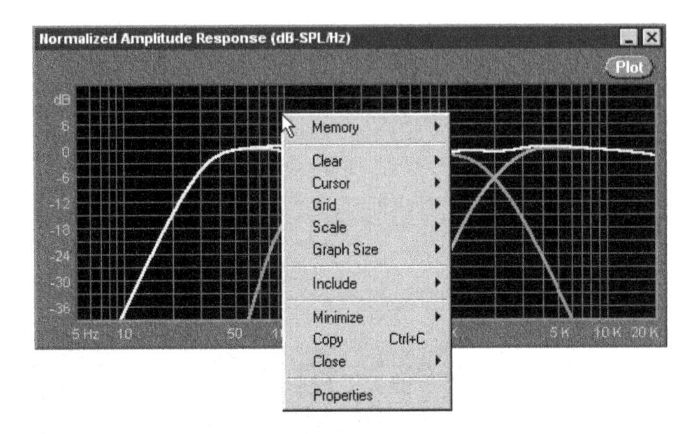

Graph Properties Window
The Graph Properties window (shown at right below) is opened by selecting the "Properties" command from the graph popup menu.

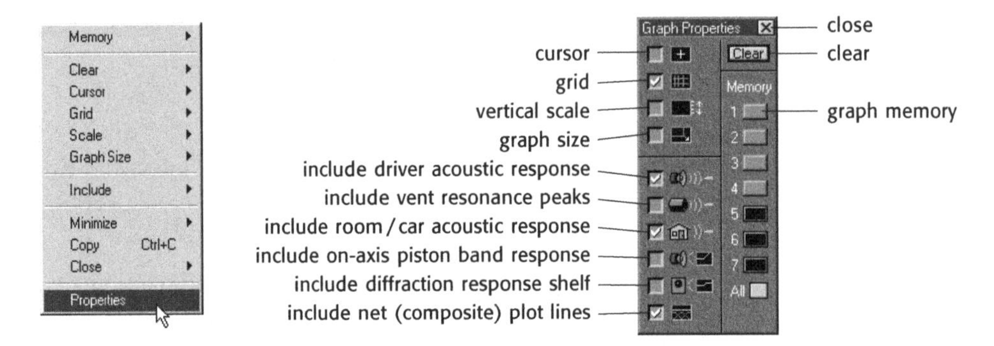

The Graph Properties window contains most of the same options as the graph popup menu. Its advantage is twofold: 1) Its many controls serve as indicators so you can quickly see how a graph is configured and which graph memories have been used; and 2) many options can be changed with a single mouse click (🖱).

Some of the settings in the Graph Properties window are specific to an individual graph. For example, the cursor can be turned on in the Normalized Amplitude Response graph and turned off in the System Impedance graph. When more than one graph is open in the "individual windows" mode, the Graph Properties window will show the settings of the selected graph window. To see the settings of a different graph window, simply click on its title bar to select it.

Each graph option is described next:

Memory Seven graph memories are provided to store a "snapshot" of the design so it can be recalled later and replotted from memory. The design parameters are stored in memory rather than the plot line data points. This enables all open graphs to be replotted when a memory is recalled–even if some of the graphs were closed when the design was originally stored into the graph memory.

To **store** the most recently plotted design into a graph memory, select the desired "Memory > Store" command from the graph popup menu (keyboard shortcut: [⇧Shift]+[F1] to [F7]). You can also hold down the [⇧Shift] key and click a memory button in the Graph Properties window. The memory button will change from black to grey to show that it is in use.

To **recall** a design from the graph memory and replot it in all open graphs, select the desired "Memory > Recall" command from the graph popup menu (keyboard shortcut: [⇧Shift]+[Ctrl]+[F1] to [F8]). You can also click on a memory button in the Graph Properties window.

To **clear** an individual graph memory, select the desired "Memory > Clear" command from the graph popup menu. You can also hold down the [Alt] key and click a memory button in the Graph Properties window. All graph memories can also be cleared with the "Clear All Memories" command of the Graph menu of the main window.

Clear Clears the plot lines from either the selected graph or all open graphs. The graph popup menu provides a separate selection for each. (Keyboard shortcuts: [Ctrl]+[X] to clear the selected graph only and [Ctrl]+[Y] to clear all graphs.) To clear only the selected graph with the Graph Properties window, click on the "Clear" button. To clear all graphs with the Graph Properties window, hold down the [⇧Shift] key and click on the "Clear" button.

5

Cursor Several cursor options are provided in the graph popup menu. The cursor in the selected graph or all graphs can be turned on/off and linked/unlinked. A keyboard shortcut is provided for turning the cursor on and off in the selected graph. Use [Ctrl]+[U] to show the cursor and [Ctrl]+[H] to hide the cursor. The Graph Properties window only allows the cursors to be turned on/off. To turn on/off the cursors for all open graphs with the Graph Properties window, hold down the [⇧Shift] key and click the Cursor button. A sample graph with a cursor is shown below:

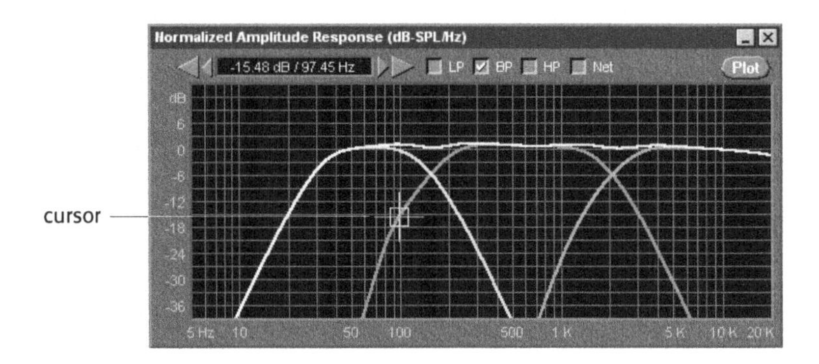

Several controls are added to the graph when the cursor is turned on (shown below):

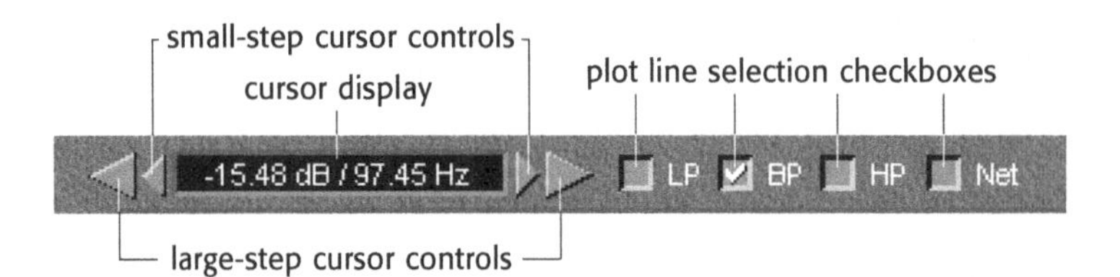

Cursor navigation buttons When the cursor is turned on, a set of Small-Step and Large-Step Cursor Controls are available to move the cursor left and right in the graph. The small-step controls move the cursor just one pixel at a time. The large-step controls move the cursor 20 pixels at a time. (A pixel is a single dot of light on the computer screen.) The cursor can also be moved with the keyboard using the [←] and [→] keys for small steps and [⇧Shift]+[←] and [⇧Shift]+[→] for large steps. A Cursor Display is located between the cursor controls to display the value and location of the cursor as it moves.

Plot line selection checkboxes Depending on the number of filters and the plot line settings, a graph can plot up to four lines for a crossover network. This includes the plot lines for the individual filters ("LP" = low-pass, "BP" = band-pass, "HP" = high-pass) and the "Net" plot line which shows the net response. The Plot Line Selection Checkboxes allow you to select which of these plot lines you want the cursor to track. The cursor will follow only the selected plot line. The plot line can also be selected from the keyboard with Ctrl+← and Ctrl+→.

As you can see below, with the many plot lines of a crossover network, a graph can quickly become cluttered. You will probably want to change the plot color each time you overlay a new plot to visually separate the plot lines. You can also turn off the "Overlay" feature in the "Graph" tab of the Preferences window. This will cause the graph to be cleared each time before plotting begins. *Note: The mini preview graph in the main window is always cleared each time it is replotted.*

The cursor normally tracks the most recent plot lines. What if the graph contains multiple plot lines from different designs like the graph below?

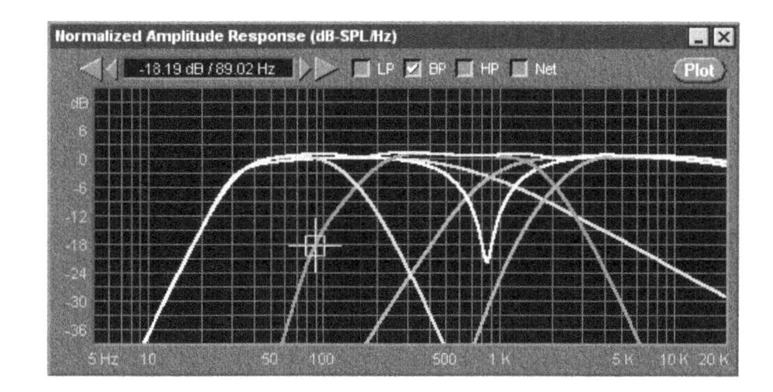

If the plot lines of multiple designs are visible in the graph as shown above, you can rotate the cursor through the last ten designs by holding down the Ctrl key and clicking one of the Large-Step Cursor Controls. This can also be accomplished from the keyboard with the ↑ and ↓ keys.

In the "individual windows" mode the cursors of all the graph windows can be linked. When this is done they will all move together so that clicking on the cursor button of one graph will cause the cursors to move in unison through all the open graphs. The link option is not available for the "single window" mode because it has only one cursor. To link the cursors, right click (🖱) on the graph and use the graph popup menu. The cursors can also be linked with default settings in the "Graph" tab of the Preferences window.

Grid If desired, the grid in one or all of the graphs can be turned off. An example with the grid turned off is shown below:

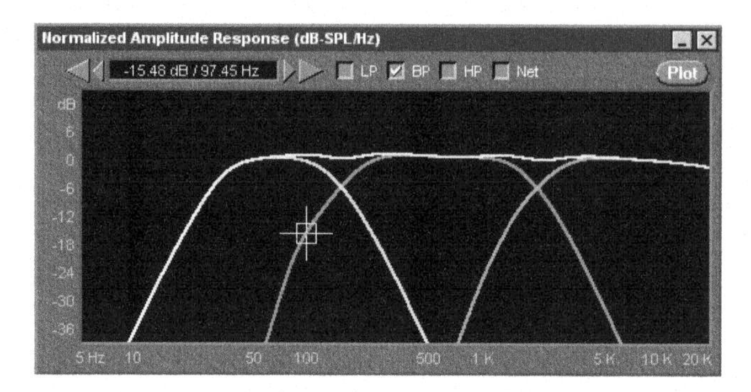

Scale The vertical scale has a "normal" and "expanded" setting. The scale used will depend on the graph. In general, the expanded selection will display a larger vertical range.

Changing the scale will cause the graph to be cleared and then replotted. Only the last ten plots will be replotted because X•over Pro has a ten-plot graph buffer. If there are more than ten plots, some will be lost. Use the graph memories to store important plots so they can be recalled even if they are not restored after the scale is changed.

Graph Size ("individual windows" mode only) The individual graph windows are available in a "normal" and "large" size. All the previous examples in this chapter have used the "normal" size graph window. A large graph window is shown below:

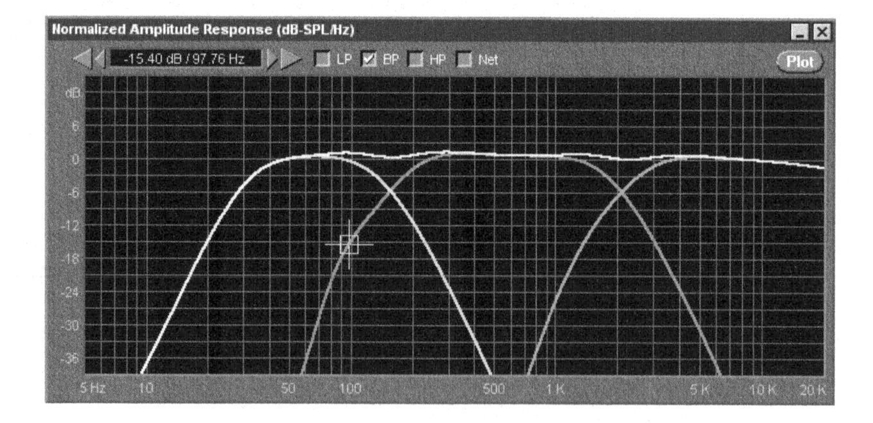

The large-size graph window does not increase the range of the vertical scale but it does add more data points, improving the detail of the plot lines. This is because the plotting functions in X•over Pro calculate a graph data point for each pixel column in the graph.

Include Six options can be included in the plotting function of several graphs. These options are global in scope. This means that turning on an "include" option will turn it on in all relevant graphs, including the mini preview graph of the main window. However, changing an "include" option will have no effect on existing plot lines—it only affects future plotting. Some of the "include" options will not be available when there is insufficient data or when an option is not relevant to the open designs. Each one is described next:

Driver Acoustic Response

If the acoustic response of the driver has been entered, it will be incorporated with the amplitude response when this option is turned on. (Keyboard shortcut: Ctrl+I.) The graphs affected by this option are: the mini preview graph and the Normalized Amplitude Response graph

Vent Resonance Peaks

The "pipe" resonance peaks created by the vent in vented and bandpass boxes will be included in the mini preview graph and all four full-size graphs when this option is turned on. The required vent data for this feature can only be imported from a BassBox Pro or BassBox Lite speaker design file—it cannot be created within X•over Pro.

Room/Car Acoustic Response

If the acoustical data for the listening space is available, it will be incorporated with the amplitude response when this option is turned on. The required acoustical data for this feature can only be imported from a BassBox Pro speaker design file—it cannot be created within X•over Pro. The graphs affected by this option are: the mini preview graph and the Normalized Amplitude Response graph.

Note: Depending on the nature of the acoustical data imported from BassBox Pro, the Driver Properties window will display a car icon for this setting if the "automotive" acoustic mode was selected when the file was created and it will display a house icon if the "architectural" acoustic mode was selected when the file was created.

On-Axis Piston Band Response

The "piston band" of a driver is related to its piston diameter. It is the frequency band where the driver maintains a constant load versus frequency. This begins at the frequency whose wavelength is equal to the circumference of the piston and it extends upward in frequency. In the piston band, the on-axis response of most drivers will begin to increase with frequency at a rate of 6 dB/octave because the beamwidth or coverage angle of the driver will gradually narrow as the frequency increases. This effect can be included in the mini preview graph and the Normalized Amplitude Response graph.

How important is it to use this option? The answer varies with each driver. If the driver's acoustic response has been entered, then this phenomenon was probably included in the acoustic measurement data. If this is the case, you should turn off this option. In the

5

past, the on-axis piston band response has been ignored because it is somewhat counteracted by the –6 dB/octave roll-off that results from the inductive rise of many voice coils. However, the two seldom cancel each other perfectly and you may want to enable this option whenever the Le parameter has been entered and disable it whenever the Le parameter is unknown.

Diffraction Response Shelf

Diffraction is the bending of sound waves as they pass near an edge or corner of a solid object. The box shape and mounting location of the driver can have an enormous effect on how severely the sound diffracts as it emanates from the driver. The "diffraction response shelf" option provides a generalized approximation of some of the diffraction effects for three box shapes: cube, square prism and optimum square prism. The required box data for this feature can only be imported from a BassBox Pro or BassBox Lite speaker design file—it cannot be created within X•over Pro. The graphs affected by this option are: the mini preview graph and the Normalized Amplitude Response graph.

Net Response (not available for "Separate Filters")

The "net" response shows the overall or composite response of the crossover network and the system. (Keyboard shortcut: [Ctrl]+[Z].) The graphs affected by the net response option are: the mini preview graph, the Normalized Amplitude Response graph and the System Impedance graph. The Phase Response and Group Delay graphs do not offer a net plot line option.

In the System Impedance graph, the net impedance response is calculated by summing the parallel impedance of each filter. In the mini preview graph and Normalized Amplitude Response graph, the net amplitude response is calculated one of two different ways depending on the setting of the "ø" (phase) checkbox in the "Graph" tab of the Preferences window. When "ø" is turned on, it is calculated by summing both the magnitude and the phase of the volume velocities of each filter. When "ø" is turned off, it is calculated by summing only the magnitudes.

Let's examine two net amplitude response examples that illustrate the difference between these two methods of calculation. Both examples use the same crossover network and drivers. The crossover point is –6 dB at 500 Hz. In the first sample at the top of the next page, both the magnitude and the phase were summed to create the net response.

The net amplitude response is represented by the white plot line. Notice the drastic effect of the phase cancellation at the crossover frequency. This is because the woofer and the tweeter are 180° out of phase at the crossover frequency. However, this method must make some important assumptions which are described in the "Assumptions" paragraph at the bottom of the next page.

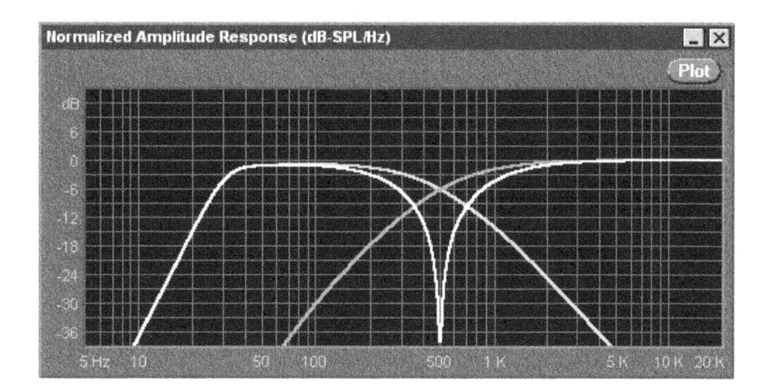

In the second example below, only the magnitude was summed to create the net amplitude response.

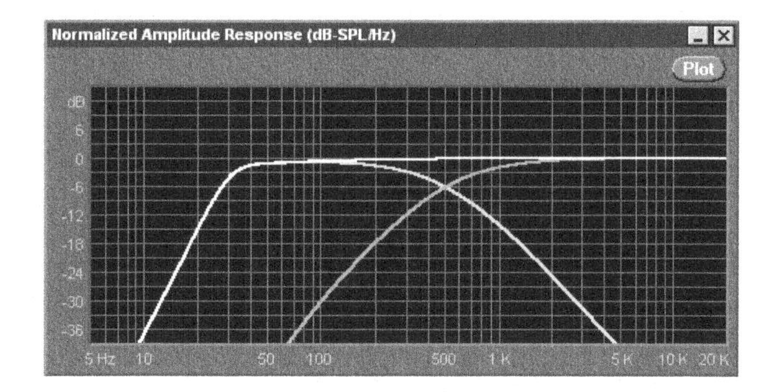

Notice in this example that no cancellation is visible. Instead, the woofer and tweeter sum to a flat response because the phase of the signals is ignored. This may sometimes be desired when driver alignment will not be attempted.

Important: Remember that all of the graphs in X•over Pro show more than just the crossover network—they also show the driver and box response. This should be apparent in two ways: 1) the high-frequency roll-off and phase shift that results from the inductive reactance of the driver's voice coil, and 2) the low-frequency roll-off and phase shift of the driver. The low-frequency response may also include the box response and phase shift for open back drivers. The result can be profound and can cause the net amplitude response to add or cancel at unexpected times when the phase of the signals is included ("ø" is turned on).

Assumptions: X•over Pro must make a few assumptions when the phase is included in the net amplitude response ("ø" is turned on). One assumption is that all drivers are carefully aligned at the crossover frequency so that the sound waves in the crossover region from each driver arrive at the listener's ear at the same time. This is a goal of

5

many high-fidelity speaker designs. However many speakers have misaligned drivers so the final amplitude and phase response can vary significantly from the model. This is one reason why it is important to prototype, test and adjust a design before finalizing it (see Chapter 7 of the *Crossover Network Designer's Guide* in this manual).

Although the graph popup menu and Graph Properties window only allow the net response to be turned on and off, it is also possible to turn on and off the individual filter plot lines. To do this use the "Graph" tab of the Preferences window.

Minimize Hides either the selected graph or all open graphs and places a graph icon on the Windows desktop. To restore the window, simply click on its icon.

Copy Creates a copy of the selected graph in the Windows clipboard so it can be pasted into another application like a word processor or page layout program. (Keyboard shortcut: Ctrl+C.) It actually "prints" to the clipboard, creating a printable version of the graph with a white background as shown below. The color of the plot lines are preserved, except for white which is switched to black. Be aware, however, that some colors like yellow may be very light.

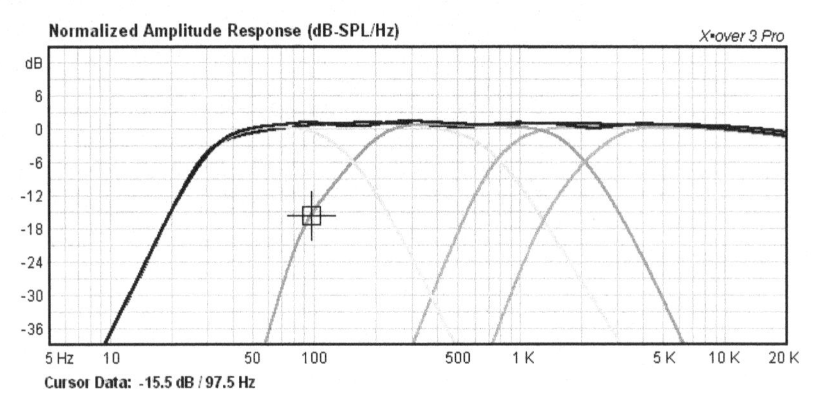

Unlike the graphs in the printouts which depict only one design, the Copy function can include the ten most recent designs that were plotted. If turned on, it also includes the cursor.

Close Closes either the selected graph or all open graphs. The graph popup menu provides a separate selection for each. You can also use the Close button in the title bar of the graph. Holding down the ⇧Shift key while clicking the Close button of a graph title bar will close all open graphs.

The Graph Properties window will be automatically closed when the last graph is closed.

Properties Opens the Graph Properties window. As long as you never explicitly close the Graph Properties window (that is, you never click on its close button), X•over Pro will automatically open and close it whenever the first graph opens and the last graph closes.

That concludes the coverage of all of the graph features in the graph popup menu and Graph Properties window. But there is one more feature that should be mentioned here...

Changing the Plot Color

Each filter has a Filter Plot Line Color indicator in the main window as shown below. Notice that the net plot line also has an indicator.

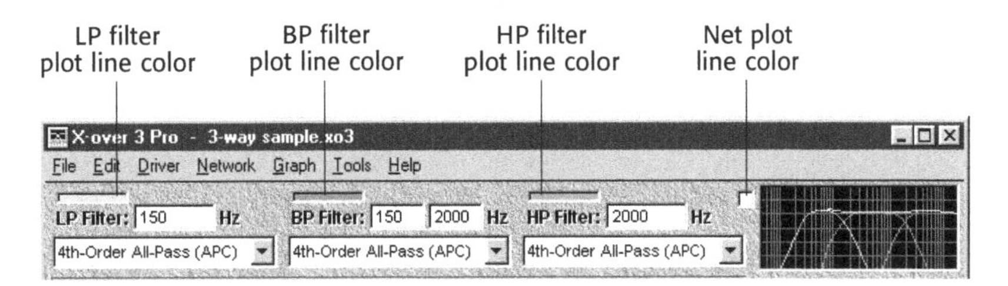

These indicators serve a dual role as both indicators and buttons. By clicking on them the color will advance one at a time through a 12-color palette with each single click. The first ten colors of the palette are fixed (red, orange, yellow, greenish-yellow, green, cyan, blue, magenta, white and light grey). But the last two colors are "custom" colors and you can make them any color you'd like with the "Graph" tab of the Preferences window. A default plot line color can also be set for each filter and the net plot line with the "Graph" tab of the Preferences window (see pages 219-220).

Changing a Filter Plot Line Color will not change the color of plot lines that already exist. It will only affect future plotting.

Next, let's examine each of the four graphs. To select a graph use the "Show Graph" command of the Graph menu as shown below:

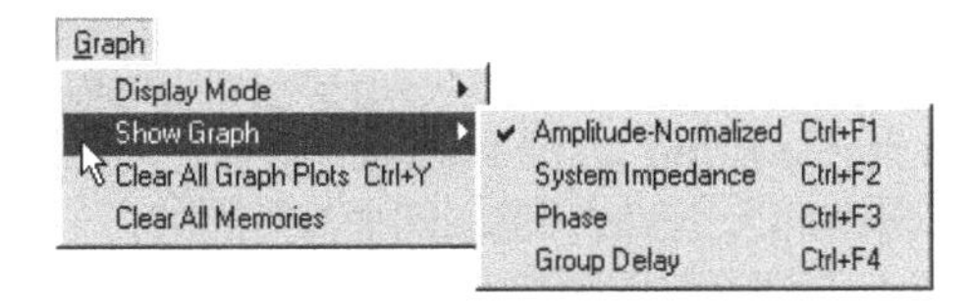

You can also use the keyboard shortcuts: Ctrl+F1 to F4.

5

5

Normalized Amplitude Response

An amplitude response graph is often called a "frequency response" graph or sometimes a "magnitude response" graph. The goal of the Normalized Amplitude Response graph in X•over Pro is to display the relative sound pressure level (SPL) differences of each part of a crossover network and the drivers attached to it.

Since "relative" differences are desired, the graph is not concerned with the absolute loudness at a particular distance and direction or with a particular input power so it is "normalized" to zero dB. But X•over Pro cannot normalize each filter separately because it needs to show the relative difference in output level between each filter. So one filter is picked from each crossover network and it is normalized. The remaining filters are scaled the same way so that level differences between them are preserved. Two-way crossover networks are normalized to the low-pass filter. Three-way crossover networks are normalized to the band-pass filter.

A special sample is shown below:

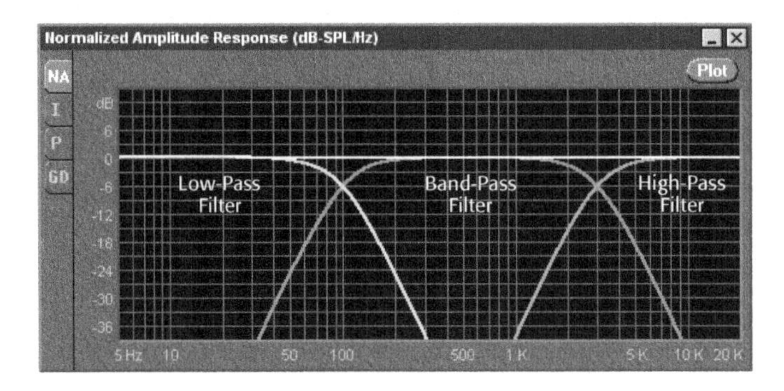

What makes this sample special (besides its DC-to-light flat response)? Extreme driver parameters were used so that the drivers would not influence the response of the network. For example, the drivers in our sample have a free air resonance (Fs) of only 1 Hz. An extreme box that is 10000000 liters in size was also used, making it very similar to an infinite baffle. The result is that the graph shows only the electrical response of the crossover network. If you allowed the X•over Pro setup program to install the sample design files when you installed the program, then you can find a copy of this design file in the "C:\Program Files\HT Audio\Designs" folder. Its file name is "Ultimate test.xo3".

The next sample uses a similar three-way 4th-order crossover network but it includes normal drivers and a vented box.

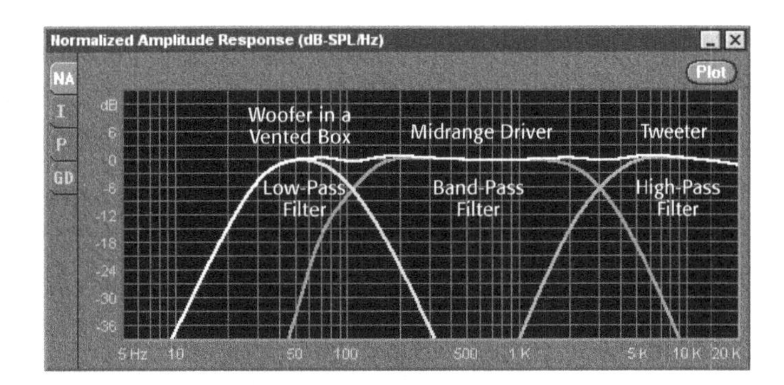

Notice that the above example appears to have two band-pass filters. Actually there is only one, the filter plot line in the middle. The left filter plot line is a low-pass filter. However, its load is the woofer in a vented box and since the graph includes the driver and box response, the low-pass filter shows the low-frequency roll-off of the woofer/box combination.

Notice also that the Normalized Amplitude Response graph has a net response option (the white plot line in the graphs) when a two-way or three-way crossover network is selected. The net amplitude response is calculated one of two different ways, depending on the setting of the "ø" (phase) checkbox in the "Graph" tab of the Preferences window. When "ø" is turned on, it is calculated by summing both the magnitude and the phase of the volume velocities of each filter. When "ø" is turned off, it is calculated by summing only the magnitudes.

The net response plot line can be turned on and off with the "Include > Net (Composite) Response" command in the graph popup menu or Graph Properties window. It can also be turned on and off with the keyboard shortcut Ctrl+Z. In addition, the individual filter plot lines can be turned on and off with the "Graph" tab of the Preferences window.

In order to understand what the curves depict in the Normalized Amplitude Response graph, it will be helpful to discuss the drivers and crossover network separately. This will be done in the following pages.

The response of a single driver

The following graph shows the response of a single woofer in a vented box. No passive network is included in this example (no crossover network, L-pad or impedance equalization network).

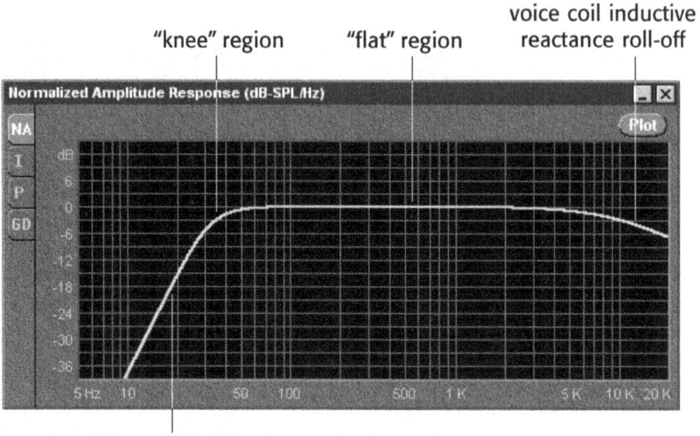

"Cutoff" region Most drivers have a low-frequency response limit—a point beyond which they can no longer reproduce sound waves efficiently. This is visible in an amplitude response graph as the portion of the plot line where the response drops rapidly. This region is known as the "cutoff" or "roll-off" region. Most designers of high-fidelity speakers try to push the low-frequency cutoff region to the lowest possible frequency. The slope of the cutoff region varies with the box type.

"Knee" region The "knee" region of the response is the portion of the plot line where the cutoff region begins. The point in the "knee" region where the response has dropped 3 dB is often labeled the "corner" frequency or "F3".

"Flat" region The "flat" region of the plot line represents the frequencies where the driver reproduces sound waves with the same loudness. For high-fidelity playback, an ideal speaker should usually have a flat response for its entire operating range or passband. Therefore, most designers try to achieve the flattest possible response.

Inductive reactance roll-off All moving coil piston drivers have inductive reactance because a coil is an inductor. This is represented with the driver parameter "Le". Unfortunately an inductor acts like a 1st-order (6 dB/octave) low-pass filter to limit the high-frequency response of audio signals flowing through it. Generally speaking, larger voice coils have greater inductance and therefore, a greater high-frequency roll-off. This is represented by the falling plot line at the high-frequency end of the graph. *Note: This is counterbalanced in some drivers by an on-axis 6 dB/octave rise in response due to the piston band narrowing of the driver's coverage angle.*

The response of a single filter

The following graph shows the response of a low-pass filter from a 100 Hz 4th-order crossover network. It has a purely resistive load in place of a driver so that only the electrical response of the filter is visible in the graph.

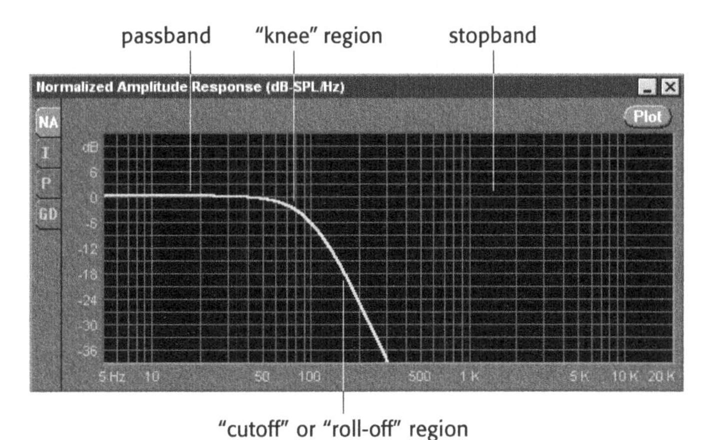

Passband The operating frequency band of the filter. It is usually measured from the –3 dB frequency in the "knee" region of the response (80 Hz in this example). The part of the passband that is visible in the above graph is 5 Hz to 80 Hz. *Note: The reason why the –3 dB point is at 80 Hz rather than 100 Hz is because this low-pass filter is from a crossover network that used a –6 dB crossover point at 100 Hz. Therefore the –6 dB frequency is 100 Hz.*

"Knee" region The portion of the plot line where the cutoff region begins. The point in the "knee" region where the response has dropped 3 dB is often labeled the "corner" frequency or "F3". *Note: Band-pass filters have two "knee" regions.*

"Cutoff" region The visible portion of the audio signal where the level drops rapidly. The slope of the cutoff region varies with the order of the filter (listed below).

1st order filter – 6 dB/octave
2nd order filter – 12 dB/octave
3rd order filter – 18 dB/octave
4th order filter – 24 dB/octave

Note: Band-pass filters have two "cutoff" regions.

Stopband The portion of the audio signal that is attenuated or "stopped" by the filter. Like the passband, it is usually measured from the –3 dB frequency in the "knee" region of the response (80 Hz in this example). The part of the stopband that is visible in the above graph is 80 Hz to 20 kHz. *Note: Band-pass filters have two "stopbands".*

The response of a speaker

The next graph depicts the system response of a complete three-way speaker. It uses a vented box and has a woofer, midrange driver, tweeter and a 3-way crossover network.

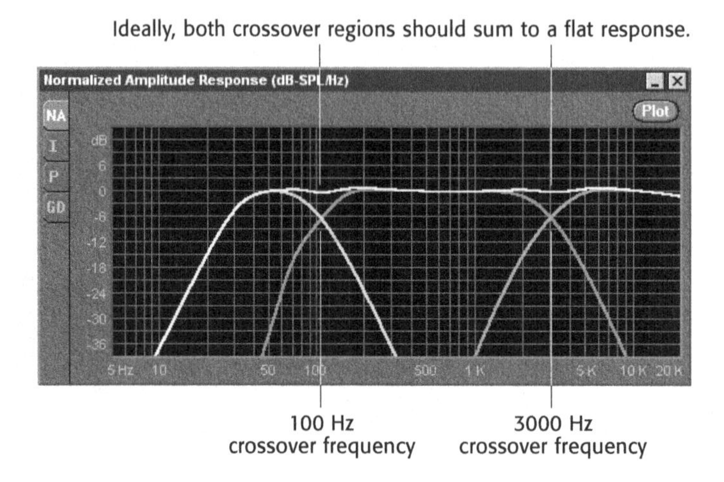

5

Since this is a three-way speaker, it has two crossover frequencies: 100 Hz and 3000 Hz. All three filters are 4th-order all-pass (APC) filters and so –6 dB crossover points were used. (For additional information about crossover network filters, see Chapter 1 of the *Crossover Network Designer's Guide* earlier in this manual.) In this case the designer wants the speaker to have a flat on-axis response so the filters and drivers should sum to a flat response in the crossover regions as shown above with the white net plot line.

Is a flat response always ideal? No. Sometimes, a non-flat response is needed to complement the acoustic environment where a speaker will be used. Musical instrument speakers, which are used for sound creation rather than reproduction, often do not have a flat response. However, most of the time, a flat response is desired for sound reproduction.

System Impedance Response

This graph shows the system impedance. This is the impedance response that the amplifier will see including the crossover network, impedance equalization networks, L-pads, drivers and box.

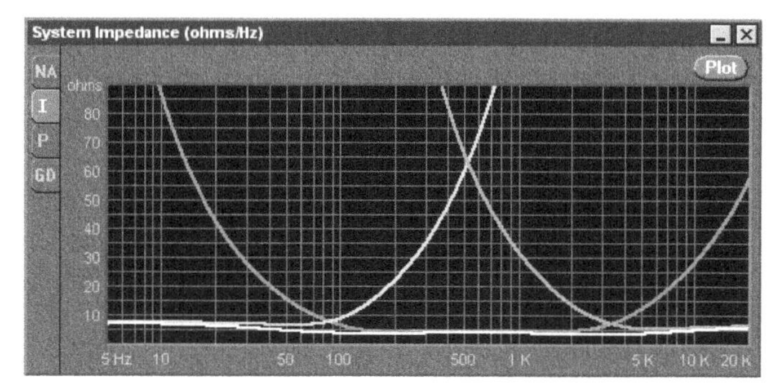

The second graph shows the same speaker without the impedance equalization networks. It underscores their value in making the impedance of each driver "look" flat to its corresponding filter. Remember that each filter "wants" a resistive load—impedance variations adversely affect performance (see Chapter 3 of the *Crossover Network Designer's Guide*).

5

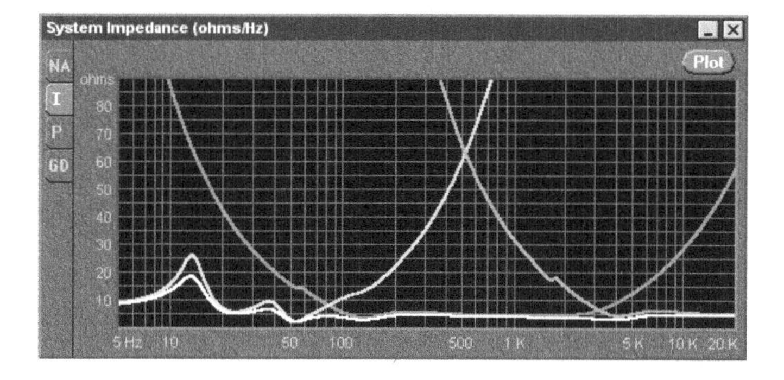

Notice that the System Impedance graph also has a net response option (the white plot line in the above graphs) when a two-way or three-way crossover network is selected. The net impedance response is calculated by adding the parallel impedances of each filter. It can be turned on and off with the "Include > Net (Composite) Response" command in the graph popup menu or Graph Properties window. It can also be turned on and off with the keyboard shortcut Ctrl+Z. In addition, the individual filter plot lines can be turned on and off with "Graph" tab of the Preferences window.

Ideally, the net impedance seen by the amplifier should be as flat as possible and it should be within the acceptable load range of the amplifier. For example, an amplifier that is designed to drive a speaker with an impedance no lower than 4 ohms may have trouble driving a speaker whose impedance response dips significantly below 4 ohms because it may try to draw too much current from the amplifier.

Whenever possible, X•over Pro uses the DC resistance value (Re) of each driver when it calculates the crossover network components. This produces better results than the nominal impedance (Z) of each driver because the actual impedance of each driver varies with frequency and drops close to the Re value. This results in a system impedance which is closer to Re rather than Z.

In many cases it is probably okay if the impedance drops a little below the minimum rated impedance of the amplifier as long as it is not in the low-frequency portion of the speaker's response and/or as long as the amplifier is not driven all the way to its full 4-ohm power rating. Consult the amplifier manufacturer for specific details about the load-handling capabilities of the amplifier.

Sometimes the band-pass filter in a three-way crossover network will have an impedance that is lower than the low-pass and high-pass filters. If the impedance of the band-pass filter is too low, see Chapter 5 ("Managing Low Band-Pass Impedance") in the *Crossover Network Designer's Guide* earlier in this manual. It contains instructions for raising the impedance of a band-pass filter.

Multiple Drivers

X•over Pro allows you to connect multiple drivers of the same type to a filter in a crossover network so it is important to understand what the System Impedance graph represents in these situations. With only one exception, the impedance response in the graph represents the net impedance of ALL the drivers in the speaker. The net impedance is calculated according to the electrical configuration setting on the "Configuration" tab of the Driver Properties window.

The one exception occurs when the "Separate" electrical configuration setting is selected for a driver. In this case the impedance response in the graph represents just ONE of the "separate" drivers. For example, a three-way speaker has two identical woofers, two identical midrange drivers and one tweeter. The two woofers are configured for "separate" wiring so each one is assumed to have its own low-pass filter. The two midrange drivers are configured for "parallel" wiring and are assumed to share a single band-pass filter. In this case, the System Impedance graph will show the impedance of a single woofer with a single low-pass filter. It will also show the combined impedance of both midrange drivers with their band-pass filter and it will show the tweeter with its high-pass filter.

Caution: It is possible for the system impedance to drop very low when multiple drivers are wired in parallel. In such circumstances, it may be necessary to wire the drivers in series so that the amplifier can handle the load.

Phase Response

This graph shows how much the sound waves emanating from each driver will lag behind the input signal. This delay is expressed as a phase angle in degrees. It is literally the difference between the phase of the input signal and the phase of the output signal. The first example below is from our "ultimate" system and reflects just the crossover network.

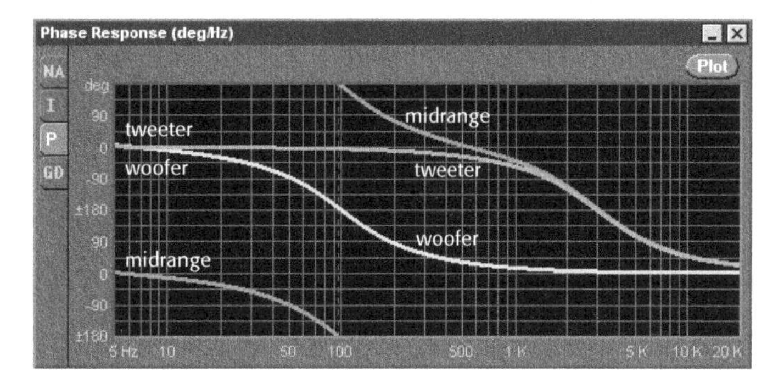

The next sample shows the same crossover network with "real-world" drivers and a vented box. Notice the greater phase lag, especially of the woofer. This is primarily due to the vented box.

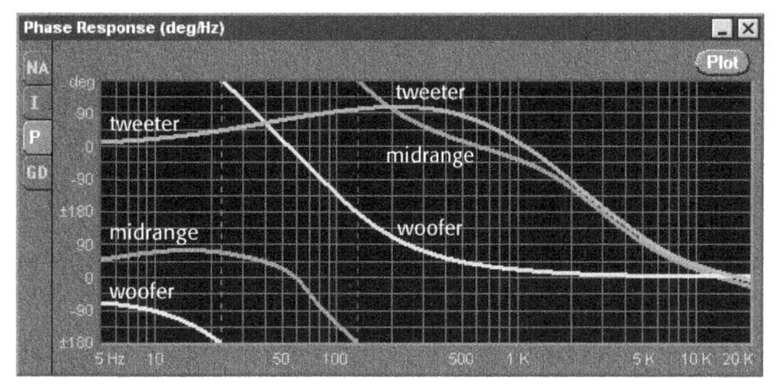

Ideally, there should be no difference in phase from one frequency to the next. This would produce flat plot lines. Unfortunately, this is not reality. Each filter of the crossover network creates a phase shift as it attenuates the sound—the higher the order of the filter, the greater the phase shift. Plus, each driver has its own inherent phase shift at both ends of its response. At the low end, the phase angle increases as the frequency decreases because of the driver's natural low-frequency roll-off. This is increased by the presence of a box. At the high end, the phase angle decreases as the frequency increases because of the inductive reactance of the driver's voice coil.

The phase of a perfect sine wave rotates 360° in one complete cycle or wavelength. If this perfect sine wave were used to drive a driver and the sound that emanated from it had no delay, the phase response would be zero degrees (0°) because the phase of both the input and output would match perfectly. If the sine wave emanating from the driver was delayed by half a wavelength, the phase response would be 180° at that frequency. In this case, the sine wave emanating from the driver would be inverted–it would be negative when the input is positive and visa versa. Both are illustrated below:

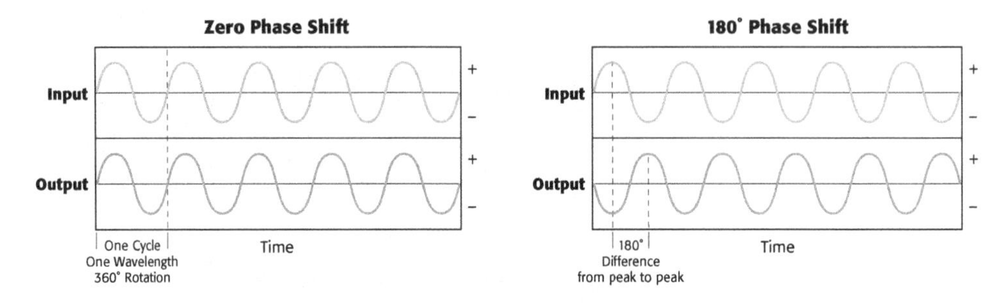

The phase response is most important in the crossover region between two adjacent drivers because their sound levels are the same at the crossover frequency. If the phase of their equal-level signals is 180° different (as shown above in the graph on the right), then the sound waves emanating from them will cancel creating a notch in the amplitude response. When this happens they are said to be "180° out of phase". This is why the phase response in the crossover region should have no sudden changes.

There is a common solution for this problem. If the phase shift between two adjacent drivers is close to 180° then inverting the polarity of one driver will bring the drivers back "in phase" at the crossover frequency. If the phase shift is close to 90° or 270° then inverting the polarity of one driver will usually not make a difference. If the phase shift is close to 0° or 360° then the polarity should not be inverted. See the "Filter Summary" section of Chapter 2 of the *Crossover Network Designer's Guide* for a description of each filter, including the phase shifts they typically produce.

Important: Remember that the graphs of X•over Pro show the system response of the speaker–not just the electrical response of the crossover network–and certain assumptions must be made about the mounting and alignment of the drivers. The reactive nature of the drivers and box will also have an effect on the phase response so the drivers will sometimes be 180° out of phase at unexpected times and visa versa. For more information see the "Polarity" section of Chapter 2 (page 130) and the "Net Response" section of Chapter 5 (pages 156-158).

A gradual phase shift is not very audible. A speaker with minimal phase shift will respond better to transient audio signals than a speaker with sudden shifts in phase. This would be important for a speaker that will be used to reproduce percussive sounds. **Tip:** The phase shift can be disregarded outside of the passband of a filter or driver.

Group Delay

This graph is similar to the Phase Response graph except that it expresses the lag of the output audio signal as a delay in milliseconds. The group delay is derived from the slope of the phase response. The first example below is from our "ultimate" system and reflects just the crossover network.

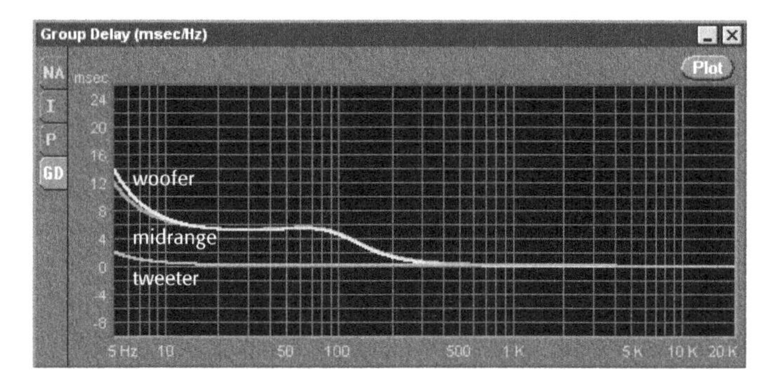

The next sample shows the same crossover network with "real-world" drivers and a vented box. Notice the greater group delay, especially of the woofer and midrange driver.

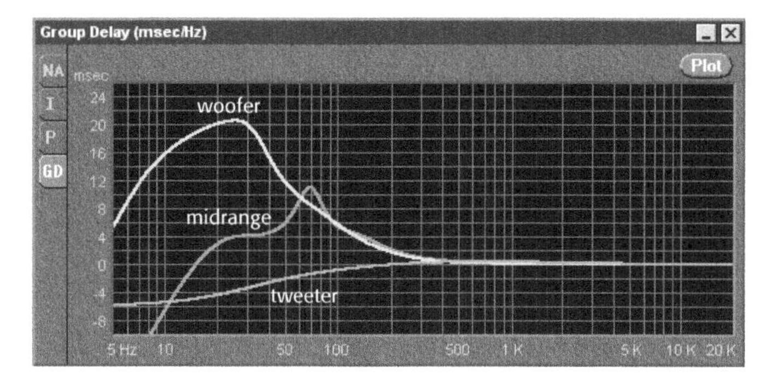

Ideally, there should be no delay from one frequency to the next within the passband of the driver and filter. This would produce flat plot lines. In reality, the group delay increases as the frequency decreases. It can be disregarded outside of the passband and a uniform delay in the passband is usually not a problem as long as it is modest or, if it is significant, the speakers are not used for live sound reinforcement or lipsync to a video image.

The most important thing to avoid is a sudden large change in group delay since this will result in poor transient response. The steeper the group delay, the worse the transient response. This can be a critical design decision for a speaker that will be used to reproduce percussive sounds.

The group delay becomes less critical at lower frequencies because their wavelengths are longer. For example, a group delay of 5 msec represents a delay in a speaker's output of 2½ wavelengths at 500 Hz. But at 50 Hz, the speaker's output will be delayed by only ¼ wavelength.

6 Saving / Opening a Design

While most computer programs run they exist in the "memory" of your computer. This is a temporary or volatile condition because the memory is erased every time the computer is shut off. In fact, X•over Pro is erased from memory each time that it is shut down with the "Quit" command of its File menu (Ctrl+Q) or Close button in its title bar.

Without a way to save your work, you would have to re-enter the driver and box parameters each time you want to resume work on a design. Fortunately, most computer programs provide a way to save and reopen your work and X•over Pro is no exception.

Saving Your Work

It is considered good practice to frequently save your work. You never know when a power outage or other "glitch" may cause your computer to "hang" or shut down. Fortunately, it is very easy to save a crossover network / filter design in X•over Pro. Simply choose the "Save Design" command in the File menu as shown below (keyboard shortcut Ctrl+S).

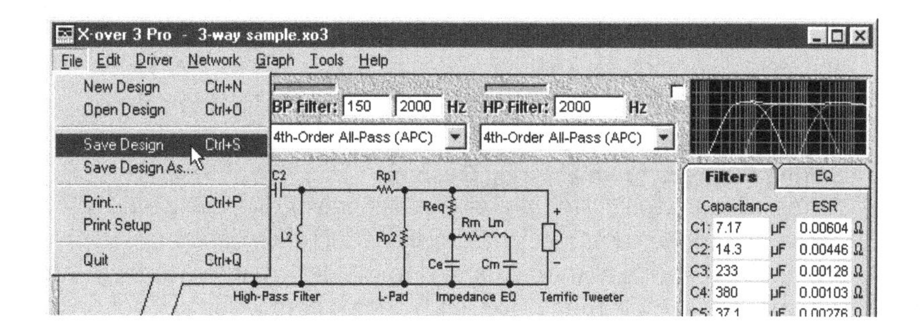

Hard Drive Basics

X•over Pro designs are saved in "files" on the "hard disk" drive of your computer. The hard disk stores the programs and files on your computer—even when the computer is turned off. Hard disks are assigned a driver letter, usually beginning with "C" ("A" and "B" are reserved for floppy disk drives). Because a hard disk can contain millions of files, the hard disk is organized into "folders" or subdirectories. Folders can contain both files and other folders.

For example, X•over Pro was stored into the "C:\Program Files\HT Audio" folder if the default settings were used during installation. This means that X•over Pro is located in the "HT Audio" folder inside the "Program Files" folder on hard driver "C". The location of a file or folder is also called its "path". For more information about your computer's hard drive, files and folders, please consult the user manuals for your computer and Microsoft Windows.

Saving a New Design

When a new design is saved for the first time, an "Enter a file name and location for the design file" dialog box (shown at right) will open to prompt you for both a location and a file name. X•over Pro supports long file names but you must always end the name with the extension ".xo3". (This is why X•over Pro design files are also referred to as "XO3" files.) The ".xo3" extension will be automatically appended if it is omitted from the end of the name.

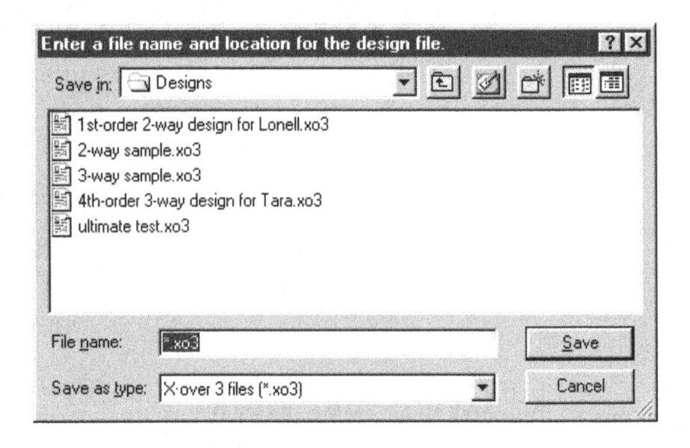

The "Enter a file name and location for the design file" dialog box is a standard Windows Open/Save dialog box. Its appearance may differ slightly depending on your Windows version. Please consult your Microsoft Windows User Manual or Windows online help if you need assistance with it.

The default location for X•over Pro design files is "C:\Program Files\HT Audio\Designs" if the default settings were used during installation of the program.

Saving Changes to an Existing Design

6

A crossover network/filter design that has previously been saved already has an X•over Pro design file on the computer's hard disk. As such, the design already has a location and file name. If you decide to make changes to the design and save them with the "Save Design" command of the File menu ([Ctrl]+[S]) shown below, you will not be prompted for a location or file name again. Instead, X•over Pro will save the design to the same location and file name as before, replacing the original file with the updated one.

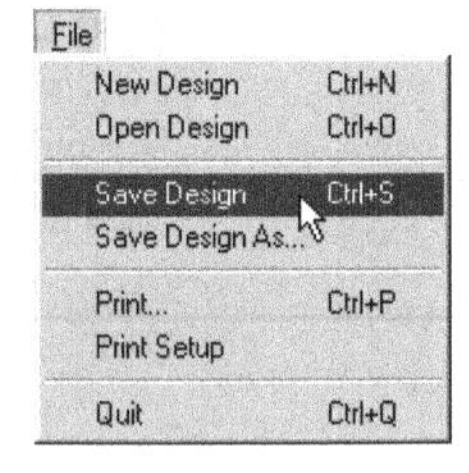

Notice that the File menu contains a second save command below the first one. The "Save Design As..." command forces the "Enter a file name and location for the design file" dialog box to appear even when an existing design already has a file name and location. Use this command when you want to save a copy of the existing design with a different file name and/or location. This creates a new copy of the design (the original copy is not changed).

Finally, when you quit X•over Pro by selecting "Quit" from the File menu ([Ctrl]+[Q]) or clicking on the Close button in the title bar of the main window, the program will first check to see if there are any changes that have not yet been saved. If there are, X•over Pro will ask if you want to save them before it shuts down.

Opening an X•over Design File

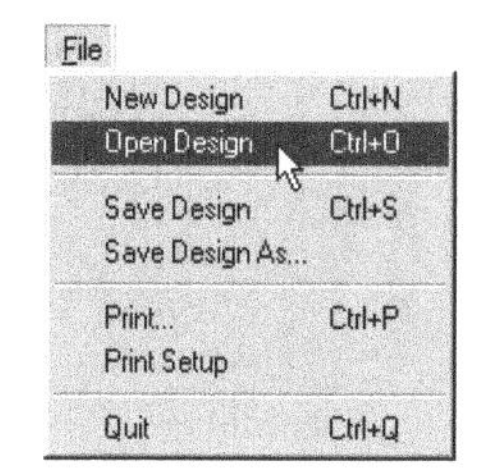

After an X•over Pro design has been saved, how do you open it again later? One way is to use the "Open Design" command in the File menu shown at right. The keyboard shortcut is Ctrl+O.

X•over Pro will open the "Open X•over Design File" dialog box to prompt you for a location and file name. This dialog is a standard Windows Open/Save dialog box and is very similar to the "Enter file name and location for the design file" dialog box shown on the previous page. Please consult your Microsoft Windows User Manual or Windows online help if assistance is needed.

Opening a Recent Design

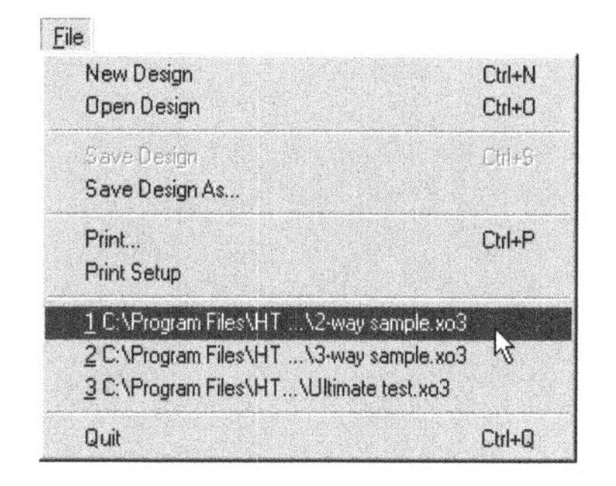

X•over Pro remembers the last four design files that were open and it adds them to the File menu. (Three recent designs are shown in the sample at right.) This provides a second way to open an X•over Pro design.

Simply select one of these files from the File menu as shown at right and it will be opened.

Auto-Starting X•over Pro by Opening a Design

There is a third way to open a design file. This method works only when X•over Pro is not running. First, locate the X•over design file that you want to open. You can do this with the Windows Explorer or you can double-click (🖱🖱) on "My Computer" on the Windows desktop and "drill" down through the folders until you locate the file. If X•over Pro was installed with the default settings, the default location for crossover network design files is "C:\Program Files\HT Audio\Designs". After locating the desired file, double-click (🖱🖱) on its icon. This will cause X•over Pro to be automatically launched and the selected design file to be automatically opened.

Importing Data from a BassBox Pro / Lite Design File

X•over Pro can open a speaker box design file from Harris Tech's box design programs BassBox Pro and BassBox Lite. This is done with the "Import" button on the Driver Properties window as shown below:

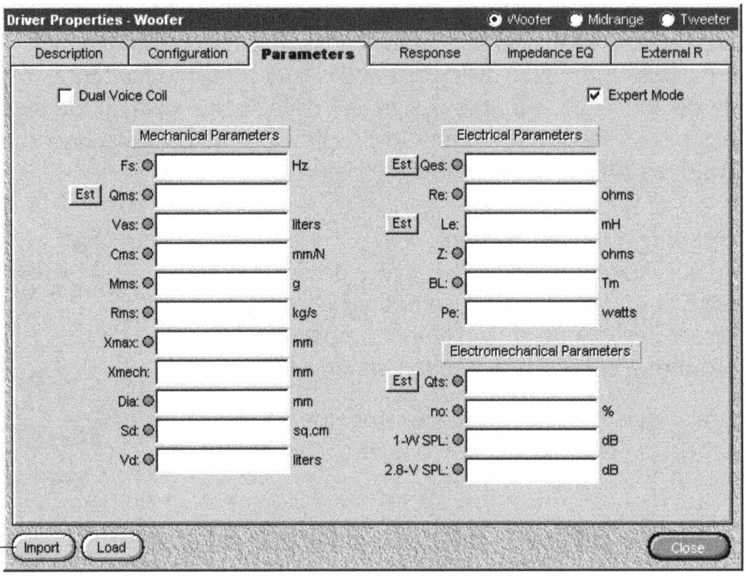

Import button

The driver parameters for an open back driver can be loaded into the "woofer" or "midrange" driver from the BassBox file. If available, the box and the acoustic response of both the driver and listening environment will also be loaded. See Chapter 2 (page 92) for more information.

File Compatibility

X•over Pro is backward compatible with earlier versions of X•over. It can open crossover network design files from X•over versions 2.1, 2.0 and 1.0.

BassBox Pro (Harris Tech's speaker box design program) can open and import a passive low-pass, band-pass or high-pass filter from a crossover network designed with X•over Pro. If present, an impedance equalization network and L-pad will be imported as well. In this way, the effects of the network can be made visible during box design. See page 209 for more information.

Finally, since X•over Pro design files contain primarily data they are not compatible with third-party programs such as word processors or page layout programs.

7 Printing a Design

To print a design, select the "Print..." command from the File menu of the main window as shown at left below. The keyboard shortcut is [Ctrl]+[P].

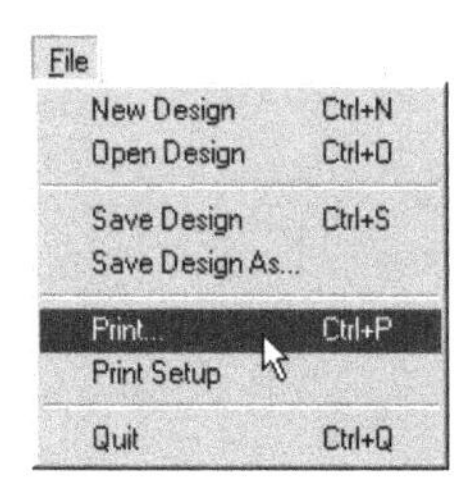

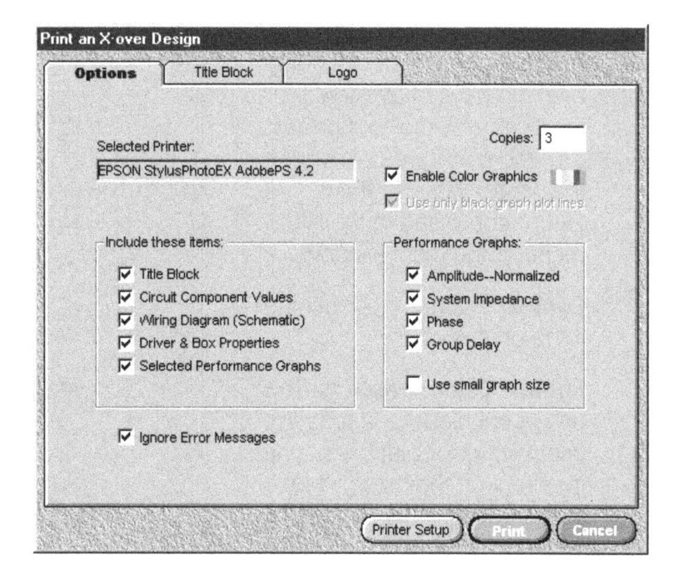

The "Print an X•over Design" window shown at right above will open. It provides a plethora of printout options which are divided into three tabs. Many of these options can be assigned default values in the Preferences window so that your choices will be selected each time X•over Pro is launched (see Chapter 11).

Before examining each tab in detail let's discuss the buttons at the bottom of the window:

Printer Setup Click the "Printer Setup" button to open the standard Windows Print Setup window so you can choose another printer and/or change the settings of the printer driver. **Important:** Make sure the printer is set to print on either a letter-size or A4-size page in portrait mode. The Print Setup window can also be opened with the "Print Setup" command in the File menu.

Print Click the "Print" button to begin printing using the selected settings.

Cancel Click the "Cancel" button to close the print window and restore the previous settings. If printing has already been started, clicking the "Cancel" button will abort printing and prevent pages that have not yet printed from being printed.

The options of each tab are described next, followed by a sample printout.

Options

The "Options" tab allows you to control the major features of a printout and identifies the selected printer.

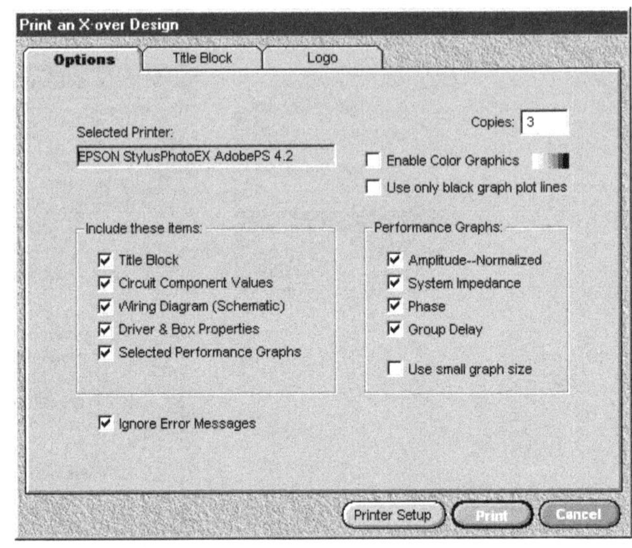

Selected Printer Displays the currently selected printer. The selected printer can be configured or changed by clicking on the "Printer Setup" button. You can also configure or change the printer with the "Print Setup" command in the File menu of the main window.

Copies The number of copies to be printed.

Enable Color Graphics This option enables the graphs to print with color plot lines. It has no effect on the logo picture. If a color logo picture is selected and a color printer is used, the logo will still print in color even if the "Enable Color Graphics" option is turned off.

Use only black graph plot lines This option controls the way graph plot lines will be printed when the "Enable Color Graphics" option is turned off. It has no effect when the "Enable Color Graphics" option is turned on.

Normally, the "Use only black graph plot lines" option should be turned off. This will allow grey plot lines to be substitued for the color graph plot lines. The shade of grey of each plot line will correspond to their original color in the on-screen graphs.

Turn on the "Use only black graph plot lines" option if your printer has trouble printing grey lines. Only black lines will be used in the printout when this option is turned on.

Include These Items There are five items that can be included in a printout. Check the ones that you want to include. They are summarized below:

> **Title Block** Includes the Title line, Designer line, Company line, Address lines, Tel/Fax line, Notes lines and the logo picture. It is printed on page one.
>
> **Circuit Component Values** Includes a list of all capacitors, inductors and resistors in the crossover network, impedance equalization networks and L-pads. It also includes general information about the crossover network such as the filter frequencies and the filter types that were used.

Wiring Diagram (Schematic) Includes a circuit diagram of the crossover network, impedance equalization networks and L-pads. It also includes configuration wiring details of the drivers.

Driver & Box Properties Includes a list of the driver and box parameters.

Selected Performance Graphs Includes the graphs checked under "Performance Graphs".

Performance Graphs Select the graphs that you want to print when the "Selected Performance Graphs" checkbox is checked under "Include these items". The "Use small graph size" option causes a smaller graph size to be used so that more graphs will fit on a page.

Ignore Error Messages Normally this option should be enabled. If you are having print problems, it can be unchecked to enable error messages to aid with troubleshooting.

Title Block

The "Title Block" tab allows you to enter information that will be printed at the top of the first page. Use it to customize the printout. For example, you can add comments for a customer or instructions for a worker.

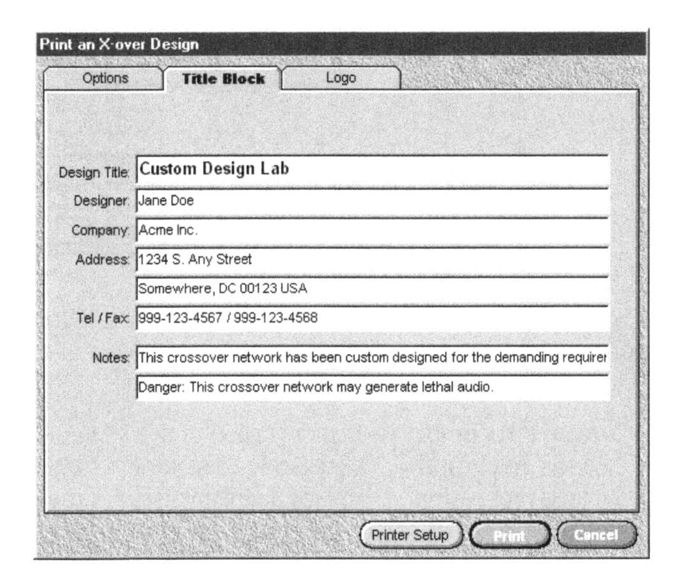

Design Title, Designer, Company, Address, Tel/Fax, Notes These lines are printed at the top of page one in the title block when the "Title Block" option is enabled on the "Options" tab. *Note: The Designer line will always be preceded with "By ".*

Logo

X•over Pro allows you to add a custom logo to your printouts. The logo prints in the title block of the first page. A smaller version also prints in the header of successive pages. The logo can be your company logo or a small picture of the box. Three different versions of the X•over Pro icon are included as default logos.

Notice below that a picture of the logo is displayed in the middle of the "Logo" tab. Whenever the mouse pointer is paused over the logo picture, the path and file name of the logo will be displayed in the balloon help.

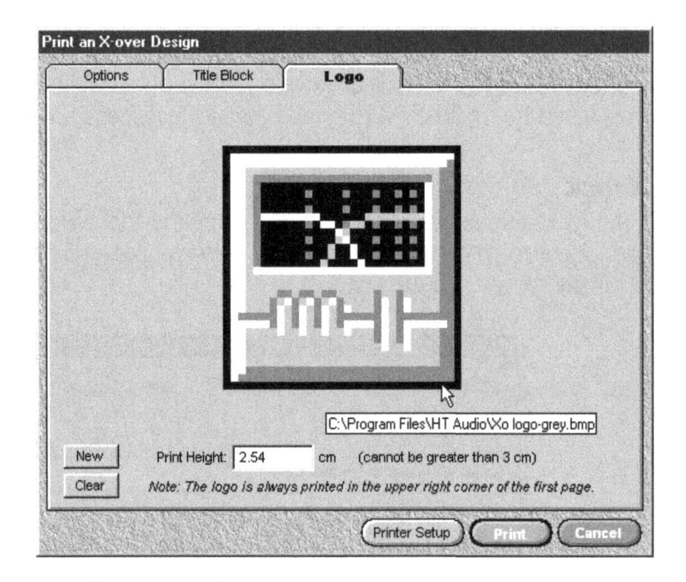

New Click this button to select a different picture for the logo. Several picture file formats are supported including: bmp, dib, gif, jpg, wmf, emf, ico and cur. Three versions of the X•over Pro icon are included with the program. They are:

Xo logo-color.bmp	a color version of the X•over Pro icon
Xo logo-grey.bmp	a greyscale version of the X•over Pro icon
Xo logo-mono.bmp	a black & white version of the X•over Pro icon

These pictures are located in the same folder as X•over Pro ("C:\Program Files\HT Audio").

Clear Clears the logo picture so that no logo will print.

Print Height Sets the height of the logo. The maximum allowable height is 1.2 inches or 30 mm. The logo picture will be proportionally scaled to this height when it is printed. Click the units label to change the units.

Sample

A sample printout follows. It includes labels to identify the various parts of each page.

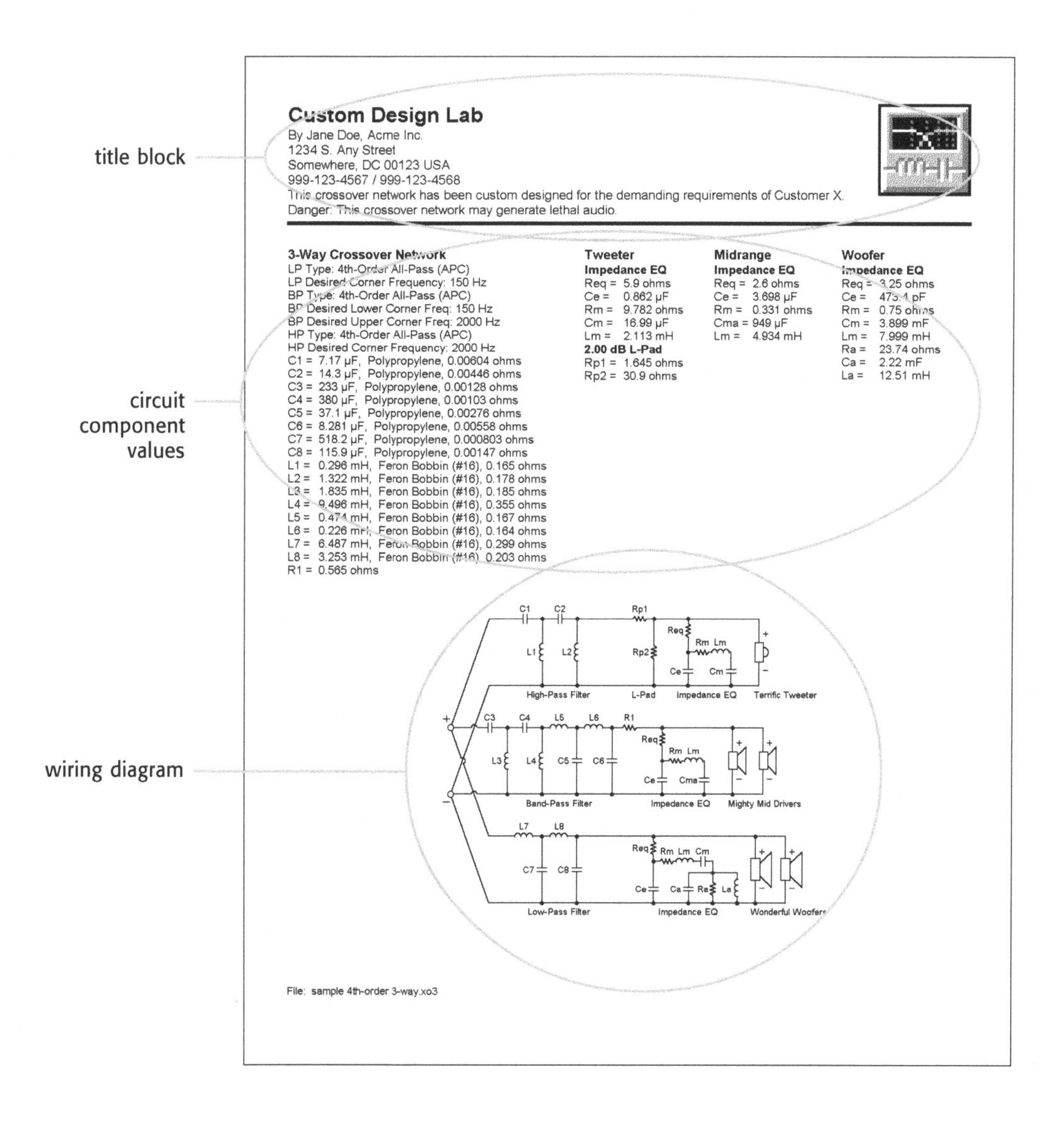

Custom Design Lab
By Jane Doe, Acme Inc.
1234 S. Any Street
Somewhere, DC 00123 USA
999-123-4567 / 999-123-4568
This crossover network has been custom designed for the demanding requirements of Customer X.
Danger: This crossover network may generate lethal audio.

3-Way Crossover Network
LP Type: 4th-Order All-Pass (APC)
LP Desired Corner Frequency: 150 Hz
BP Type: 4th-Order All-Pass (APC)
BP Desired Lower Corner Freq: 150 Hz
BP Desired Upper Corner Freq: 2000 Hz
HP Type: 4th-Order All-Pass (APC)
HP Desired Corner Frequency: 2000 Hz
C1 = 7.17 µF, Polypropylene, 0.00604 ohms
C2 = 14.3 µF, Polypropylene, 0.00446 ohms
C3 = 233 µF, Polypropylene, 0.00128 ohms
C4 = 380 µF, Polypropylene, 0.00103 ohms
C5 = 37.1 µF, Polypropylene, 0.00276 ohms
C6 = 8.281 µF, Polypropylene, 0.00558 ohms
C7 = 518.2 µF, Polypropylene, 0.000803 ohms
C8 = 115.9 µF, Polypropylene, 0.00147 ohms
L1 = 0.296 mH, Feron Bobbin (#16), 0.165 ohms
L2 = 1.322 mH, Feron Bobbin (#16), 0.178 ohms
L3 = 1.835 mH, Feron Bobbin (#16), 0.185 ohms
L4 = 9.496 mH, Feron Bobbin (#16), 0.355 ohms
L5 = 0.474 mH, Feron Bobbin (#16), 0.167 ohms
L6 = 0.226 mH, Feron Bobbin (#16), 0.164 ohms
L7 = 6.487 mH, Feron Bobbin (#16), 0.299 ohms
L8 = 3.253 mH, Feron Bobbin (#16), 0.203 ohms
R1 = 0.565 ohms

Tweeter
Impedance EQ
Req = 5.9 ohms
Ce = 0.862 µF
Rm = 9.782 ohms
Cm = 16.99 µF
Lm = 2.113 mH
2.00 dB L-Pad
Rp1 = 1.645 ohms
Rp2 = 30.9 ohms

Midrange
Impedance EQ
Req = 2.6 ohms
Ce = 3.698 µF
Rm = 0.331 ohms
Cma = 949 µF
Lm = 4.934 mH

Woofer
Impedance EQ
Req = 3.25 ohms
Ce = 473.4 pF
Rm = 0.75 ohms
Cm = 3.899 mF
Lm = 7.999 mH
Ra = 23.74 ohms
Ca = 2.22 mF
La = 12.51 mH

File: sample 4th-order 3-way.xo3

Notice below and on the next page that the full-size graphs are used. No more than three full-size graphs will fit per page. To print all four graphs requires two pages.

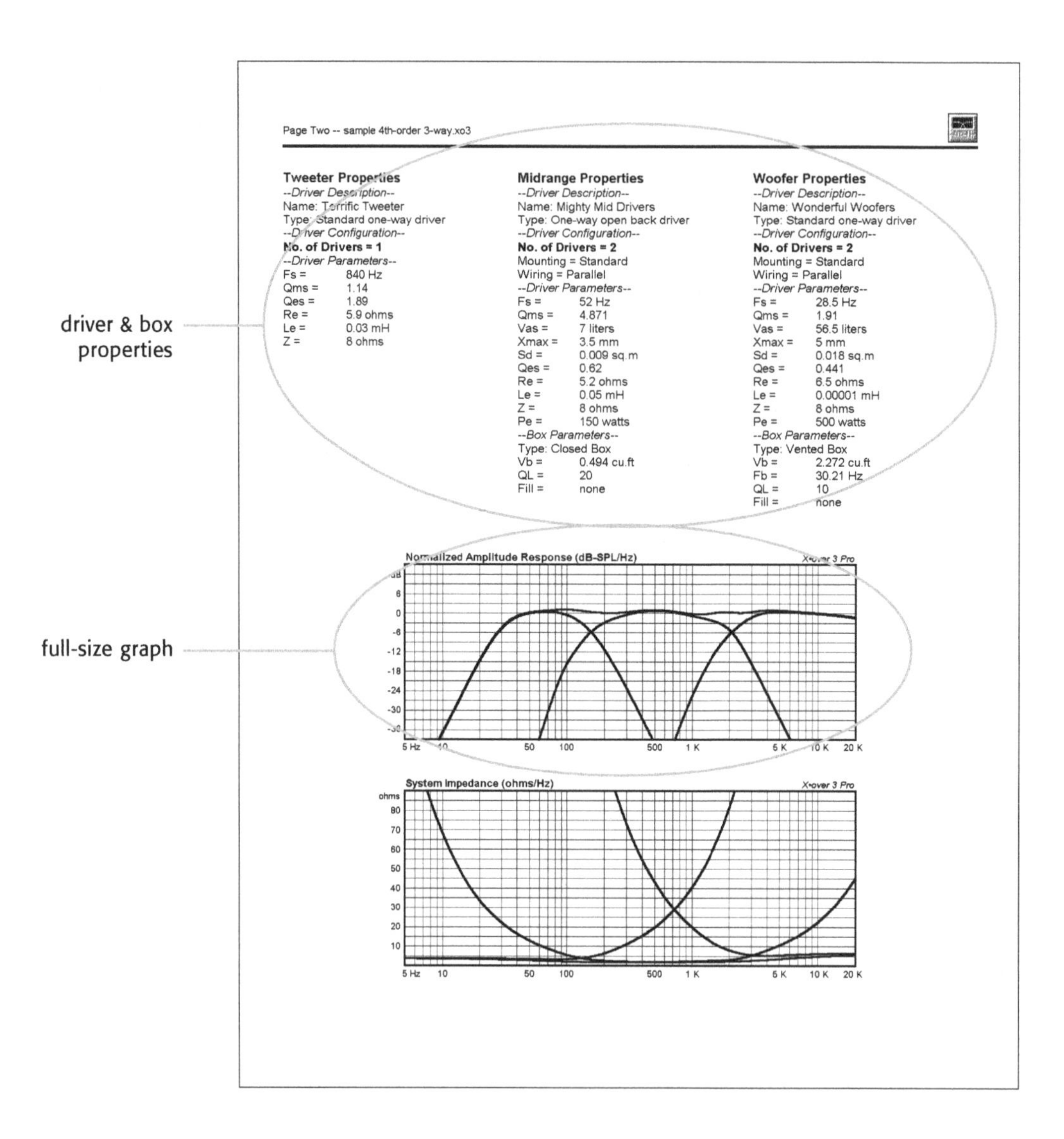

Page Two -- sample 4th-order 3-way.xo3

Tweeter Properties
--Driver Description--
Name: Terrific Tweeter
Type: Standard one-way driver
--Driver Configuration--
No. of Drivers = 1
--Driver Parameters--
Fs = 840 Hz
Qms = 1.14
Qes = 1.89
Re = 5.9 ohms
Le = 0.03 mH
Z = 8 ohms

Midrange Properties
--Driver Description--
Name: Mighty Mid Drivers
Type: One-way open back driver
--Driver Configuration--
No. of Drivers = 2
Mounting = Standard
Wiring = Parallel
--Driver Parameters--
Fs = 52 Hz
Qms = 4.871
Vas = 7 liters
Xmax = 3.5 mm
Sd = 0.009 sq.m
Qes = 0.62
Re = 5.2 ohms
Le = 0.05 mH
Z = 8 ohms
Pe = 150 watts
--Box Parameters--
Type: Closed Box
Vb = 0.494 cu.ft
QL = 20
Fill = none

Woofer Properties
--Driver Description--
Name: Wonderful Woofers
Type: Standard one-way driver
--Driver Configuration--
No. of Drivers = 2
Mounting = Standard
Wiring = Parallel
--Driver Parameters--
Fs = 28.5 Hz
Qms = 1.91
Vas = 56.5 liters
Xmax = 5 mm
Sd = 0.018 sq.m
Qes = 0.441
Re = 6.5 ohms
Le = 0.00001 mH
Z = 8 ohms
Pe = 500 watts
--Box Parameters--
Type: Vented Box
Vb = 2.272 cu.ft
Fb = 30.21 Hz
QL = 10
Fill = none

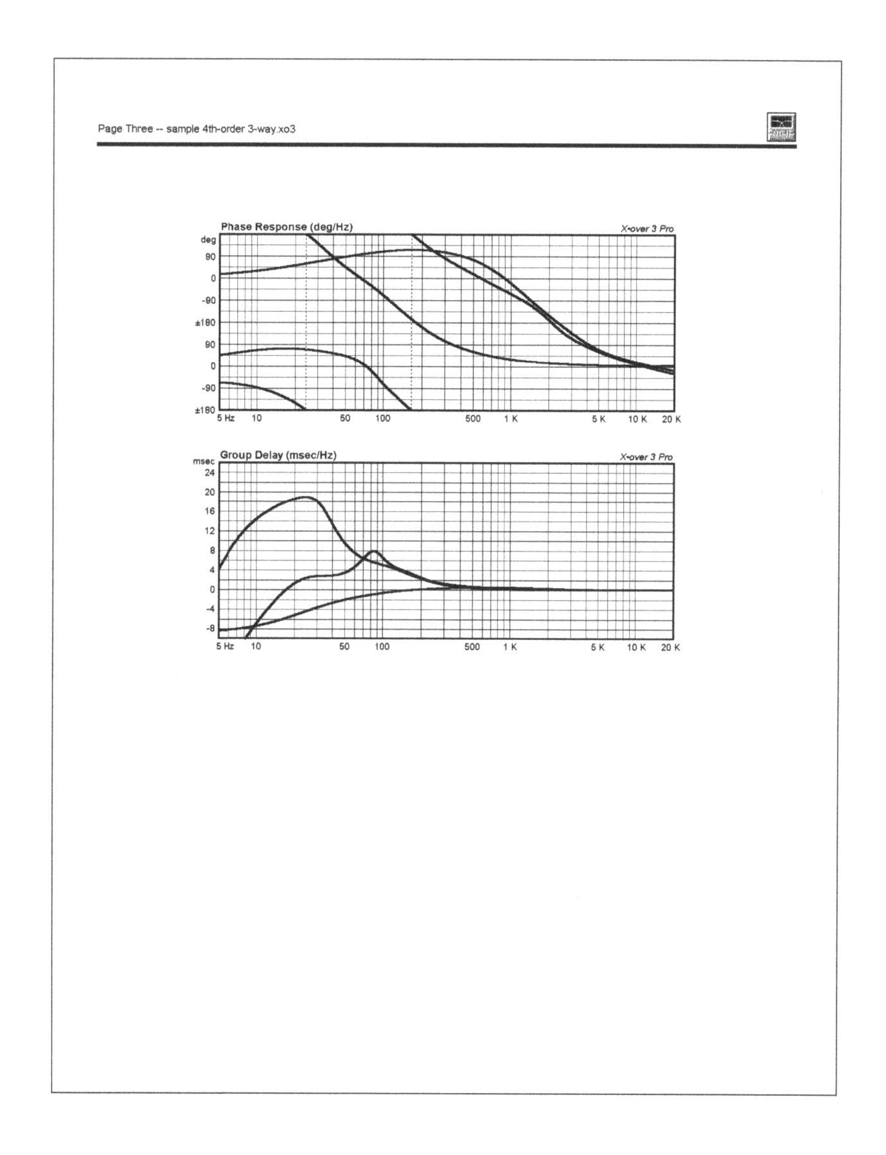
Page Three -- sample 4th-order 3-way.xo3
Phase Response (deg/Hz)
X•over 3 Pro
deg
90
0
-90
±180
90
0
-90
±180
5 Hz
10
50
100
500
1 K
5 K
10 K
20 K
Group Delay (msec/Hz)
X•over 3 Pro
msec
24
20
16
12
8
4
0
-4
-8
5 Hz
10
50
100
500
1 K
5 K
10 K
20 K

7

The sample below shows an alternate to Page 2 and 3. It shows the graphs when the "Fit all graphs on one page" option is turned on.

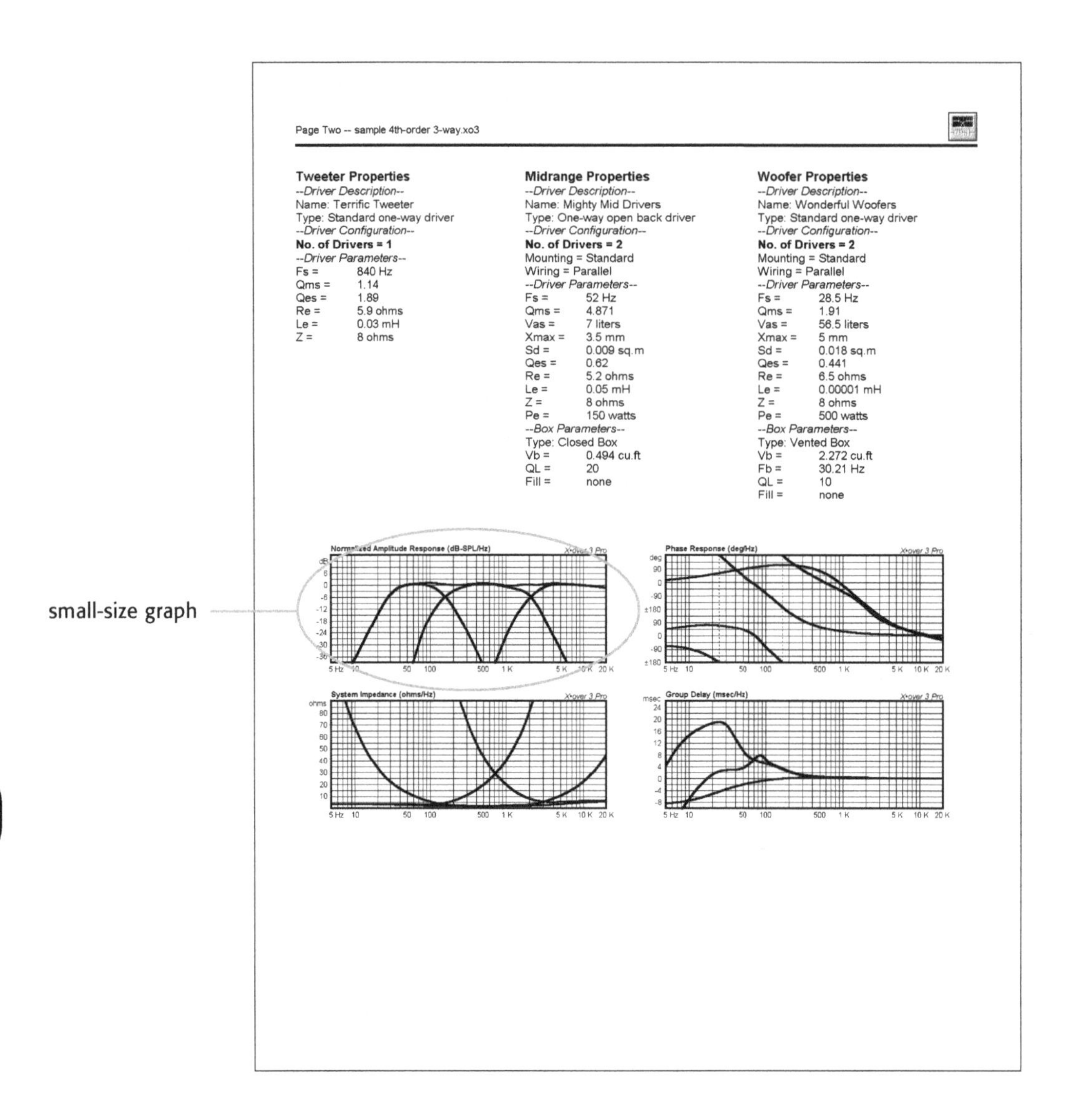

Page Two -- sample 4th-order 3-way.xo3

Tweeter Properties
--Driver Description--
Name: Terrific Tweeter
Type: Standard one-way driver
--Driver Configuration--
No. of Drivers = 1
--Driver Parameters--

Fs =	840 Hz
Qms =	1.14
Qes =	1.89
Re =	5.9 ohms
Le =	0.03 mH
Z =	8 ohms

Midrange Properties
--Driver Description--
Name: Mighty Mid Drivers
Type: One-way open back driver
--Driver Configuration--
No. of Drivers = 2
Mounting = Standard
Wiring = Parallel
--Driver Parameters--

Fs =	52 Hz
Qms =	4.871
Vas =	7 liters
Xmax =	3.5 mm
Sd =	0.009 sq.m
Qes =	0.62
Re =	5.2 ohms
Le =	0.05 mH
Z =	8 ohms
Pe =	150 watts

--Box Parameters--
Type: Closed Box

Vb =	0.494 cu.ft
QL =	20
Fill =	none

Woofer Properties
--Driver Description--
Name: Wonderful Woofers
Type: Standard one-way driver
--Driver Configuration--
No. of Drivers = 2
Mounting = Standard
Wiring = Parallel
--Driver Parameters--

Fs =	28.5 Hz
Qms =	1.91
Vas =	56.5 liters
Xmax =	5 mm
Sd =	0.018 sq.m
Qes =	0.441
Re =	6.5 ohms
Le =	0.00001 mH
Z =	8 ohms
Pe =	500 watts

--Box Parameters--
Type: Vented Box

Vb =	2.272 cu.ft
Fb =	30.21 Hz
QL =	10
Fill =	none

Exporting & Printing a Design

There is more than one way to print a crossover network design. X•over Pro provides three ways to export portions of the design to other programs via the Windows clipboard. Once the design is exported to another program, you can use its unique print features. The first two methods are in the Edit menu as shown below:

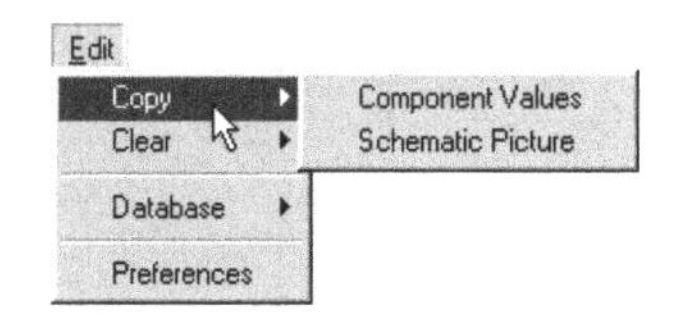

Copying Component Values

Select the "Copy > Component Values" command from the Edit menu to copy a text list of the capacitors, inductors and resistors used in the crossover network, impedance equalization networks and L-pads. Also included are the filter frequencies and filter types. The list is copied to the Windows clipboard as text and can be "pasted" into most Windows word processing and page layout programs. A sample is shown below:

3-Way Crossover Network
Low-Pass (LP) Filter: 1 required
Type: 4th-Order All-Pass (APC)
Desired Corner Frequency: 150 Hz
Band-Pass (BP) Filter: 1 required
Type: 4th-Order All-Pass (APC)
Desired Lower Corner Freq: 150 Hz
Desired Upper Corner Freq: 2000 Hz
High-Pass (HP) Filter: 1 required
Type: 4th-Order All-Pass (APC)
Desired Corner Frequency: 2000 Hz
C1 = 7.17 μF, Polypropylene, 0.00604 ohms
C2 = 14.3 μF, Polypropylene, 0.00446 ohms
C3 = 233 μF, Polypropylene, 0.00128 ohms
C4 = 380 μF, Polypropylene, 0.00103 ohms
C5 = 37.1 μF, Polypropylene, 0.00276 ohms
C6 = 8.281 μF, Polypropylene, 0.00558 ohms
C7 = 518.2 μF, Polypropylene, 0.000803 ohms
C8 = 115.9 μF, Polypropylene, 0.00147 ohms
L1 = 0.296 mH, Air Core(#16), 0.165 ohms
L2 = 1.322 mH, Air Core(#16), 0.178 ohms
L3 = 1.835 mH, Air Core(#16), 0.185 ohms
L4 = 9.496 mH, Air Core(#16), 0.355 ohms
L5 = 0.474 mH, Air Core(#16), 0.167 ohms
L6 = 0.226 mH, Air Core(#16), 0.164 ohms
L7 = 6.487 mH, Air Core(#16), 0.299 ohms
L8 = 3.253 mH, Air Core(#16), 0.203 ohms
R1 = 0.565 ohms

Tweeter
Impedance EQ
Req = 5.9 ohms
Ce = 0.862 μF
Rm = 9.782 ohms
Cm = 16.99 μF
Lm = 2.113 mH
L-Pad: 2.00 dB
Rp1 = 1.645 ohms
Rp2 = 30.9 ohms

Midrange
Impedance EQ
Req = 2.6 ohms
Ce = 3.698 μF
Rm = 0.331 ohms
Cma = 949 μF
Lm = 4.934 mH

Woofer
Impedance EQ
Req = 3.25 ohms
Ce = 473.4 pF
Rm = 0.75 ohms
Cm = 3.899 mF
Lm = 7.999 mH
Ra = 23.74 ohms
Ca = 2.22 mF
La = 12.51 mH

Note: The list is really only one column wide. It is displayed here in two columns to conserve space.

Copying a Schematic Picture

Select the "Copy > Schematic Picture" command from the Edit menu to copy a black and white picture of the schematic. The schematic is copied to the Windows clipboard as a bit-map picture and can be "pasted" into most Windows paint programs, image editors, word processing and page layout programs. A sample is shown below:

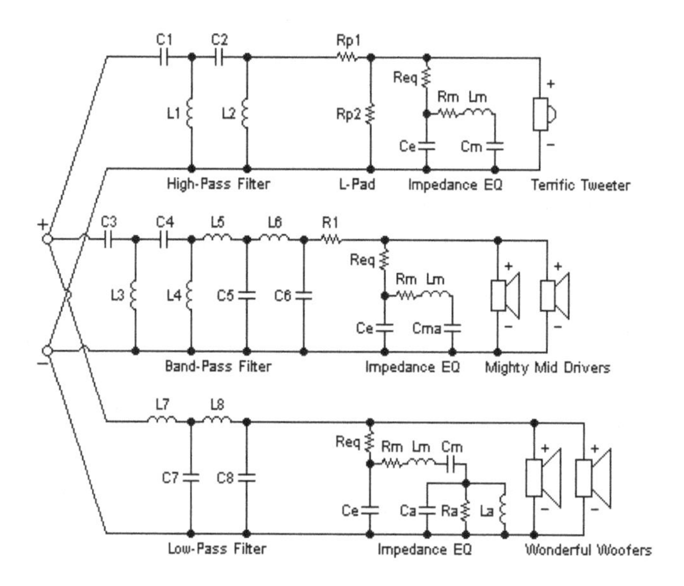

Copying a Graph Picture

Select the "Copy" command from the graph pop-up menu (Ctrl+C) to copy a color picture of a graph with a white background. It is copied to the Windows clipboard as a bitmap picture and can be "pasted" into most Windows paint programs, image editors, word processing and page layout programs. Right click on the graph to open the pop-up menu:

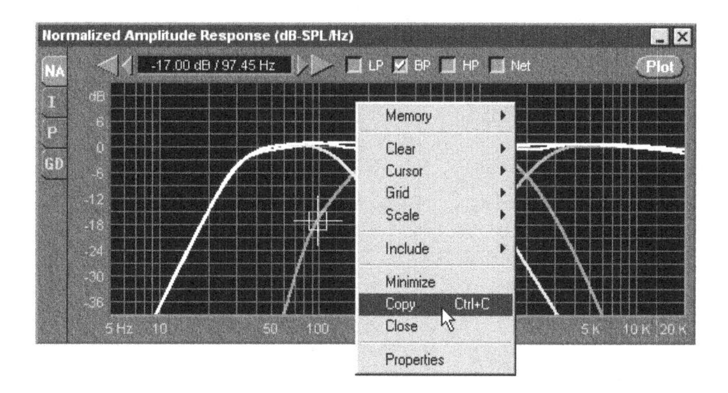

When the graph picture is created, white plot lines will be switched to black. Other colors will be unchanged. This may cause a problem because some colors like yellow are very light. You may want to change the plot colors and replot the graph before you copy it to the Windows clipboard.

Unlike the graphs in the printouts which depict only one design, graphs that are copied to the clipboard will show the ten most recent plots and, if it is turned on, the cursor. A sample is shown below:

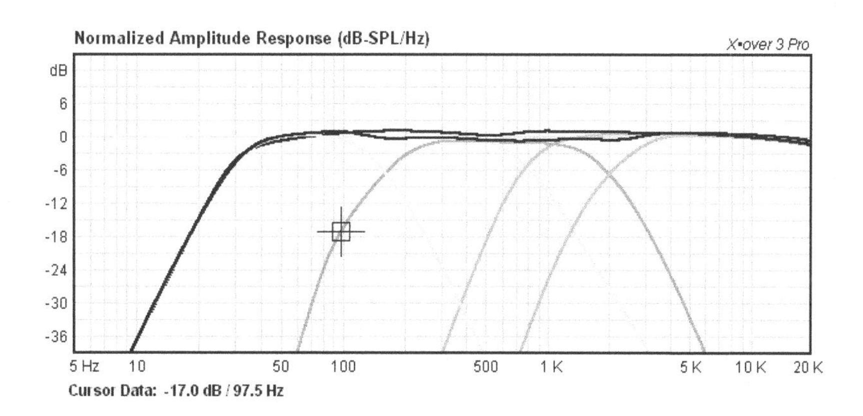

8 Clearing a Design

X•over Pro provides several commands for clearing all or part of a crossover network design. This chapter describes each one.

Clearing an Entire Design

To clear all elements of a design including all information about the drivers, box, crossover network, impedance equalization filters and L-pads, select the "New Design" command from the File menu (Ctrl+N) as shown below:

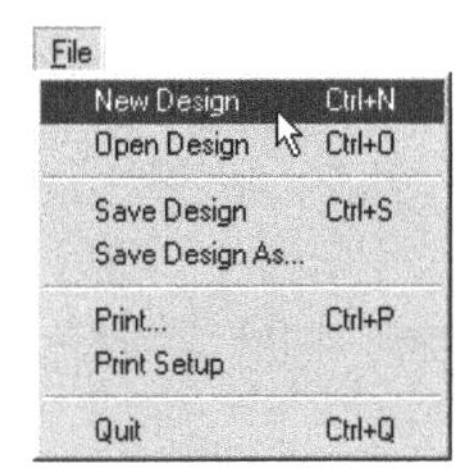

Clearing All Circuits

To clear all circuits in a design including the crossover network or filters, impedance equalization networks and L-pads, select the "Clear > All Circuits" command from the Edit menu as shown below.

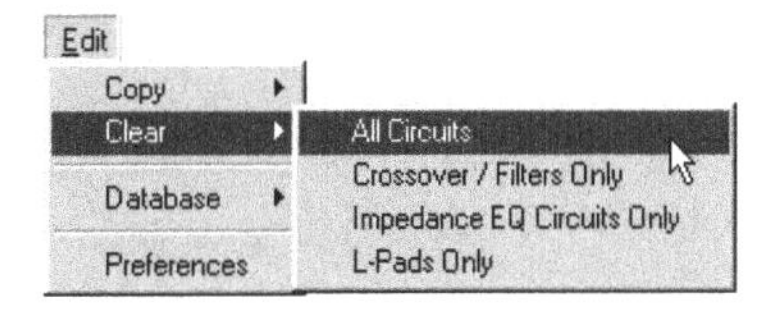

This command will not clear the driver or box information.

Clear Crossover Network/Filters Only

To clear only the crossover network or filter, select the "Clear > Crossover / Filters Only" command from the Edit menu as shown below.

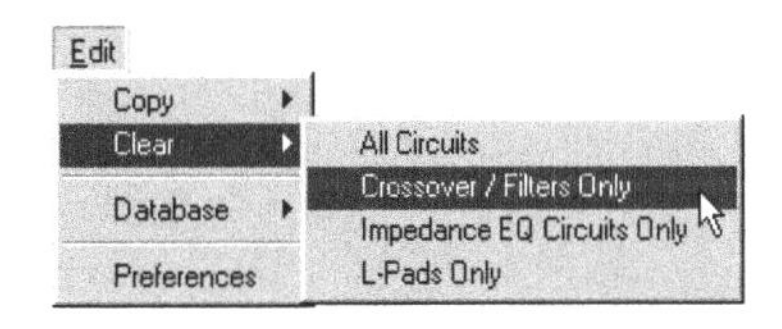

This command will not clear the driver or box information, impedance equalization networks or L-pads.

8

Clear Impedance EQ Circuits Only

To clear only the impedance equalization networks, select the "Clear > Impedance EQ Circuits Only" command from the Edit menu as shown below:

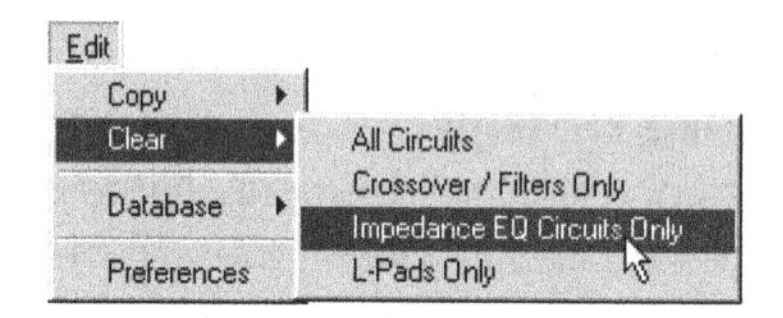

This command will not clear the driver or box information, crossover network or filters, or L-pads.

Clear L-Pads Only

To clear only the L-pads, select the "Clear > L-Pads Only" command from the Edit menu as shown below:

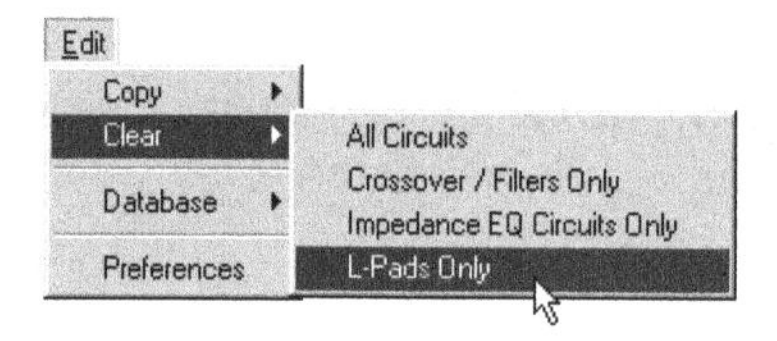

This command will not clear the driver or box information, crossover network or filters, or impedance equalization networks.

9 Editing the Driver Database

Chapter 2 (pages 82-91) discussed how to search the driver database to locate a driver and load its parameters into the program. Chapter 2 (page 81) also mentioned the "Add this Driver to Database" button of the Driver Properties window which allows the user to add a new driver to the database. This chapter will show the best way to add a driver to the database as well as how to edit and delete the ones that are already there.

To edit the driver database, select the "Database > Edit Driver Data" command in the Edit menu of the main window as shown below. The keyboard shortcut is Ctrl+W.

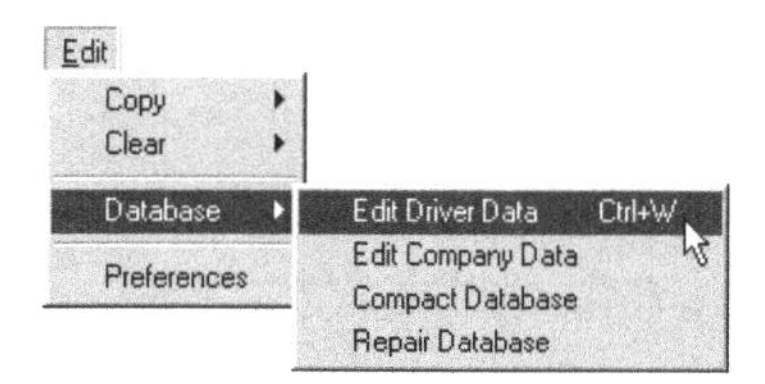

Two windows will open (shown below): the Edit Database Driver Data window and the Driver Locator window. The Edit Database Driver Data window is used to edit the driver data while the Driver Locator is used to navigate the database.

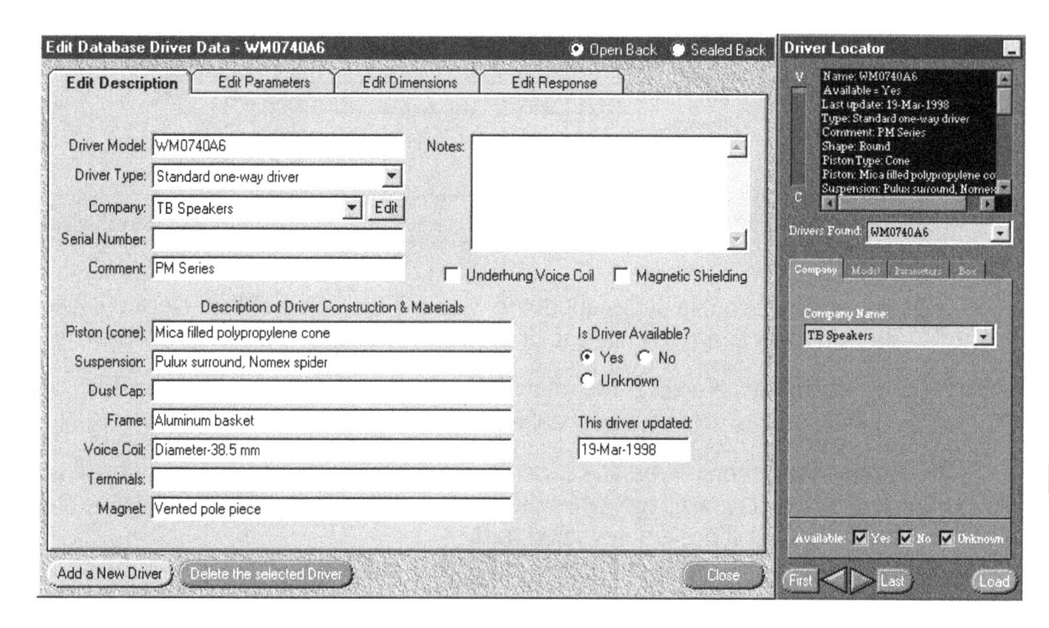

Since the operation of the Driver Locator window was covered in Chapter 3, only a couple of quick points will be provided here. First, the Driver Locator will not have a Close button in its title bar because it is automatically closed when the Edit Database Driver Data window is closed. Instead it will have a Minimize button which, in addition to serving its normal function, can also be used to switch the size of the Driver Locator window. This is done by holding down the ⇧Shift key when you click on the Minimize button.

Second, the default company which is selected when the Driver Locator first opens is set in the Preferences window (see Chapter 11). If no default company is selected in the Preferences window, the Driver Locator will simply open to the first driver in the database.

Now on to the Edit Database Driver Data window...

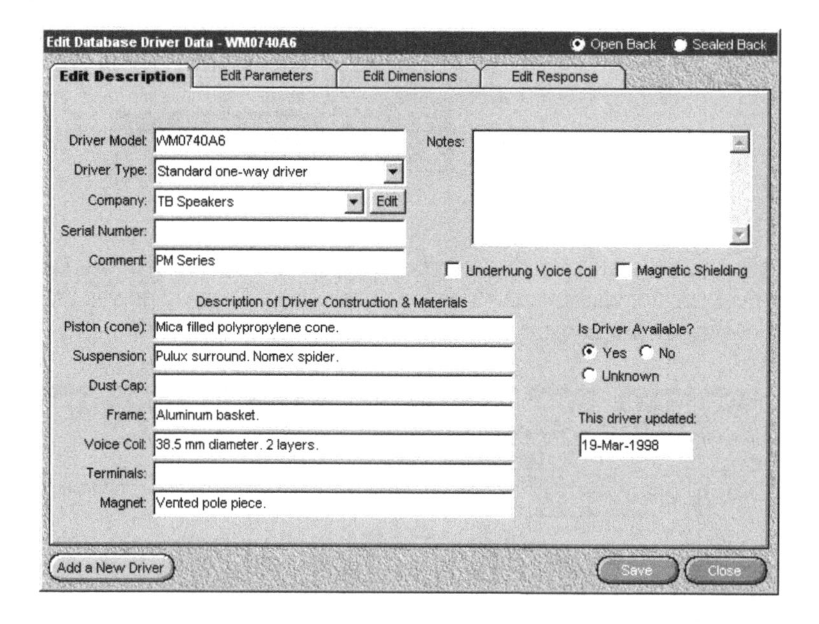

Let's notice a couple things about the window. The name of the selected driver is added to the window's title in the title bar. In the above example, driver "WM0740A6" is selected.

Number values are never rounded in the Edit Database Driver Data window. This prevents rounding errors when entering data into the database.

Notice the buttons in the title bar and on the bottom of the window in the illustration above and on the previous page. Which buttons are available is determined by the state of the data. Their functions are described next:

Open Back, Sealed Back Remember that X•over Pro divides the drivers in the driver database into two groups: drivers with open backs and drivers with sealed backs (see pages 41-42). The driver database allows only one type to be viewed at a time. Use the "Open Back" and "Sealed Back" buttons to change the selection.

Add a New Driver This button is always available to clear the current driver and begin a new one. It allows some of the information to be repeated from the previous driver. This is described in the "Adding a New Driver" section below.

Delete the Selected Driver (shown on page 189) This button is only available when there are no unsaved changes. It disappears as soon as a driver is edited and reappears when its changes are saved. Its function is to delete a single driver from the database

Save This button is only available when there are unsaved changes. It saves the changes to the database record occupied by the selected driver.

Close This button is always available. It closes both the Edit Database Driver Data window and the Driver Locator window.

Important: X•over Pro does not consider the editing of the driver database to be a casual affair. It will ask you to confirm the commands that, if executed, will change the database. Deleting a driver, saving a new driver and saving changes to an existing driver are three of the actions that will always require confirmation.

Adding a New Driver

To add a new driver follow these steps:

1 Click on the "Add a New Driver" button. If you don't begin by clicking this button, the program will assume that you are editing the selected driver. After clicking the "Add a New Driver" button, the Add New Driver window shown below will open.

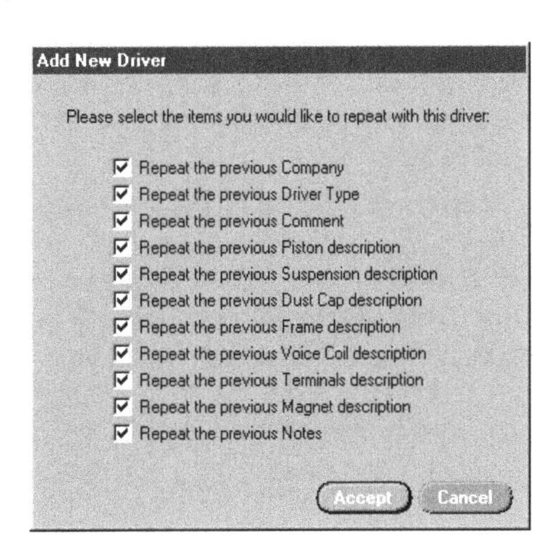

9

This window allows you to repeat the company of the previous driver and many of the descriptions. This can save time when entering several similar drivers.

2 Enter the requested information in the "Enter Description" tab. Much of this information can be repeated from the previous driver as described in Step 1. The required Description parameters are listed below.

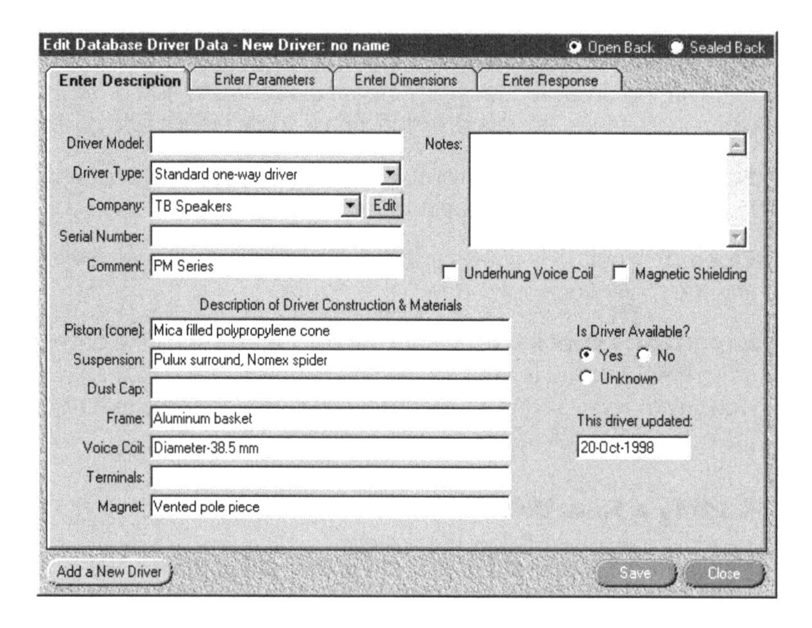

Driver Model The model name can be up to 30 characters long. A company cannot have more than one of each model name.

Driver Type Select one of the four choices: standard one-way driver, two-way coaxial driver, two-way coincident driver or three-way driver. See Chapter 2 (pages 94-95) for a description of each.

Company Each driver must be linked to an existing company in the database. This should be the company that is considered the "manufacturer" of the driver—not the distributor. The "Company" drop-down list contains the names of all these companies. If the new driver is manufactured by a company that is not in the database, the company must be added before the driver can be added. To enter a new company (or edit an existing one) click on the "Edit" button to the right of the "Company" drop-down list. This is described in more detail in the "Editing Companies" section later in this chapter.

Is Driver Available? Select the answer to this question. If you do not know, select "unknown".

This driver updated The current date will be added to this field automatically. However, it is best to enter the date of the driver information. From time to time, manufacturers make changes ("improvements") to their drivers. By entering the date of the data, you will know when newer information is available. Some manufacturer's spec sheets will have a date on them. If not, try contacting the manufacturer to find out the date for the specs.

Please enter the date in the international form: dd-mmm-yyyy. For example enter July 27, 2004 as 27-Jul-2004.

This concludes the required information for the "Enter Description" tab. The remaining descriptions and notes of this tab are optional but highly recommended. See Chapter 2 (pages 94-96) if you'd like more information.

3 Using the "Enter Parameters" tab enter the Thiele-Small and electromechanical parameters for the driver.

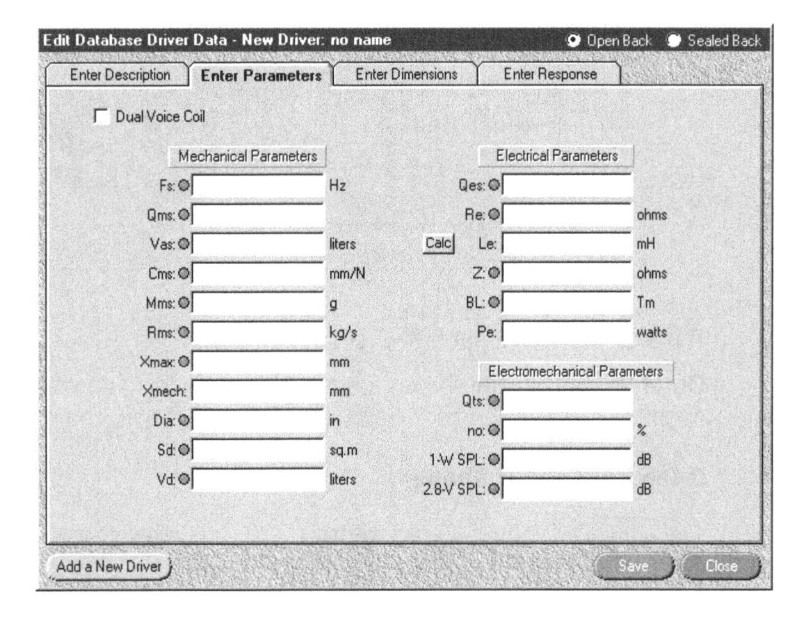

Unlike the "Parameters" tab in the Driver Properties window which has two modes (normal and expert), this "Enter Parameters" tab is always in "expert mode" to help identify data problems before the parameters are saved to the database.

A minimum set of parameters must be entered or the driver cannot be added to the database. The minimum parameters include Fs, Vas (open back drivers only) and either Qms, Qes or Qts. A model name and valid company name are also required. See

9

Chapter 2 (pages 104-112) for a description of each parameter. Before entering a parameter, remember to select the desired units by clicking on its unit label.

4 Using the "Enter Dimensions" tab select a shape and piston type for the driver and then enter its dimensions.

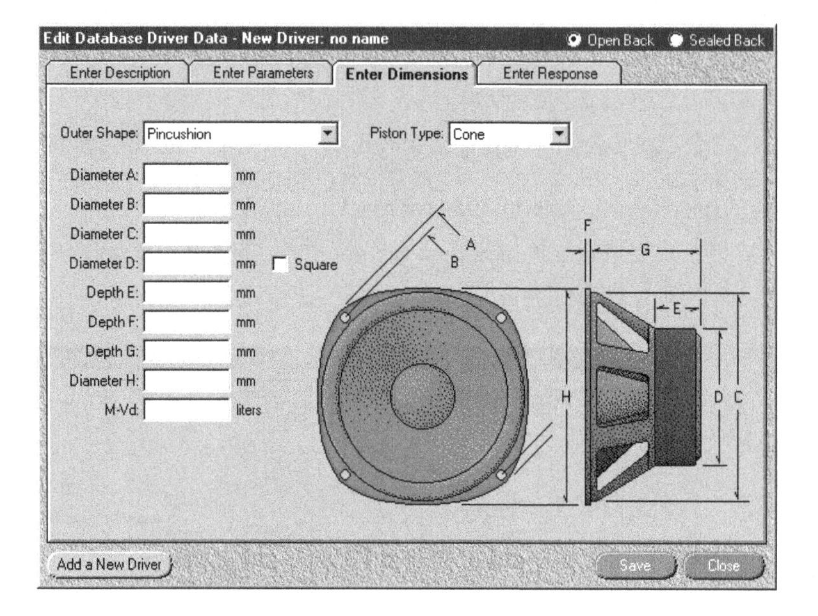

This step is optional but highly recommended.

Outer Shape Several shapes are available depending on whether the driver has an open back or a sealed back. Samples are shown below and on the next page:

Open Back Outer Shapes:

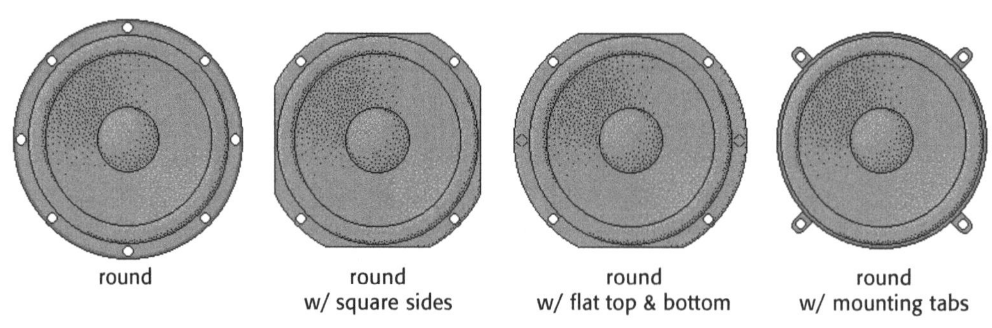

Open Back Outer Shapes continued...

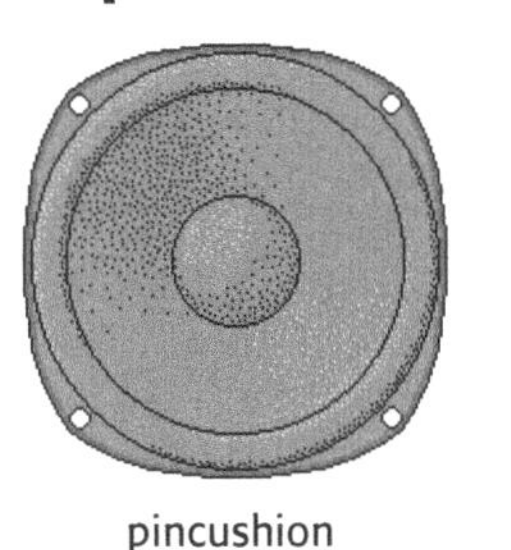
pincushion

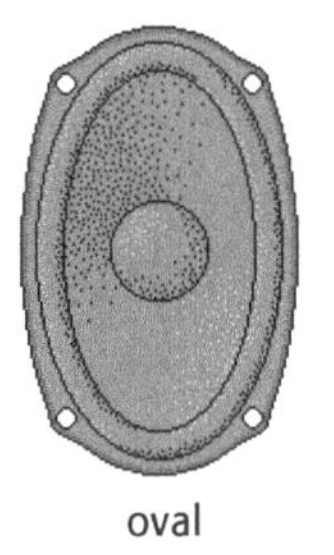
oval

square

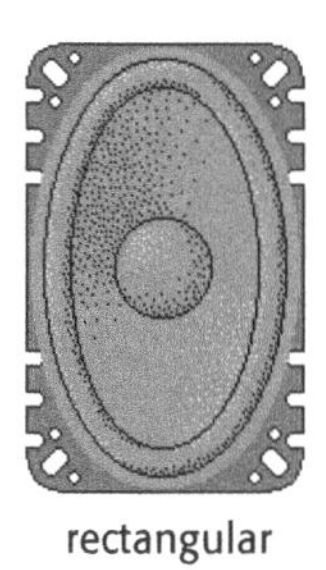
rectangular

Sealed Back Outer Shapes:

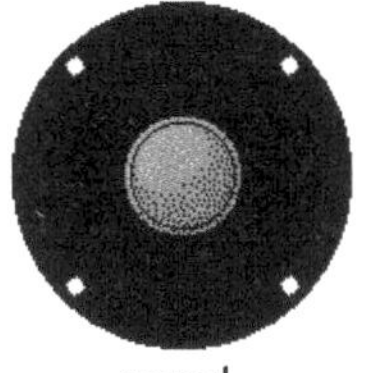
round

round w/ square sides

round w/ one flat side

round w/ two flat sides

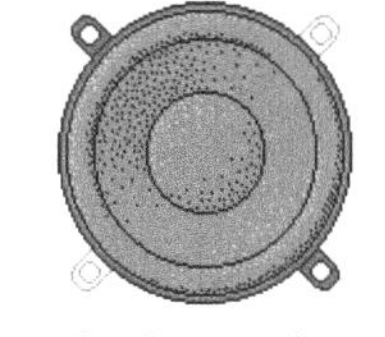
round w/ mounting tabs

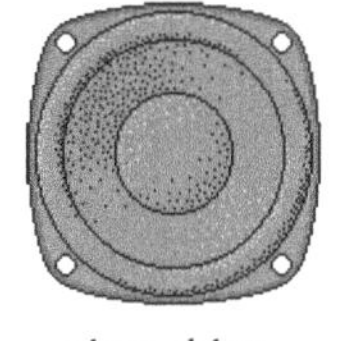
pincushion

square

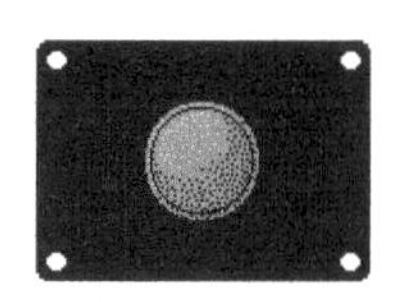
rectangular

Piston Type Several piston types are available depending on whether the driver has an open back or a sealed back and depending on the "Outer Shape". Samples are shown below and on the next page:

Open Back Piston Types:

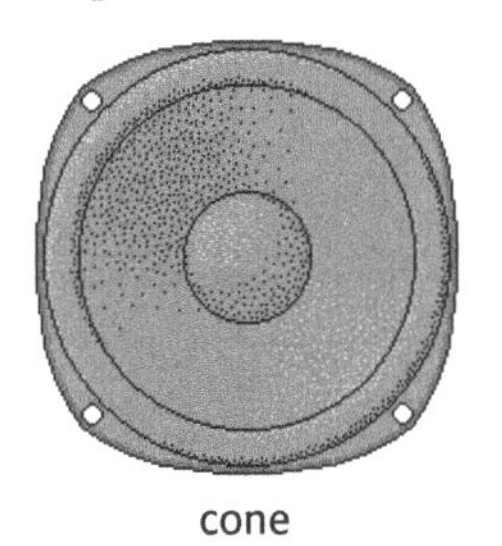
cone

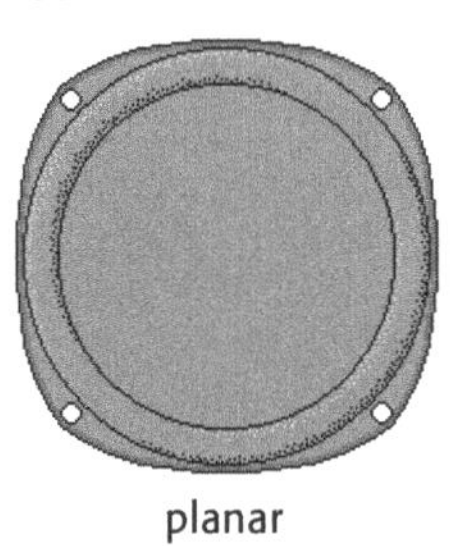
planar

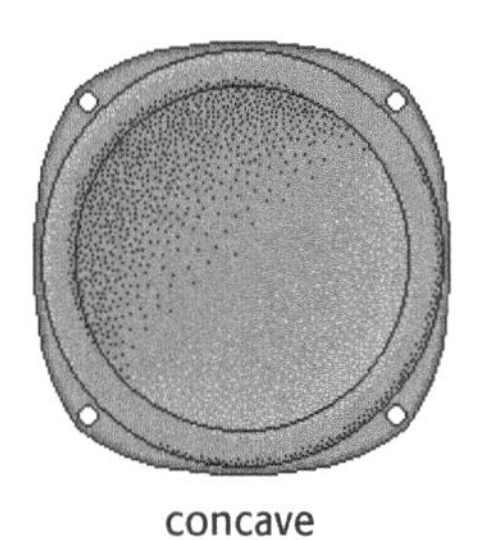
concave

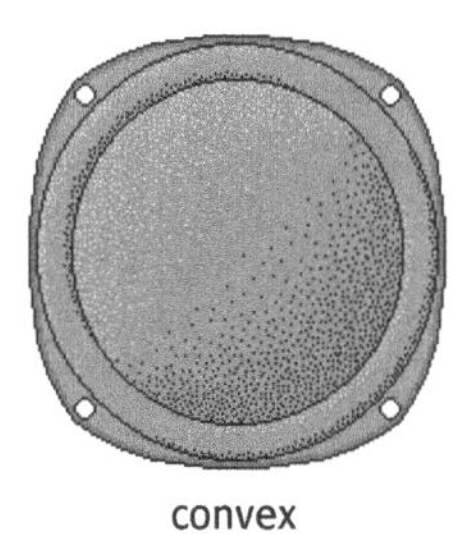
convex

Sealed Back Piston Types:

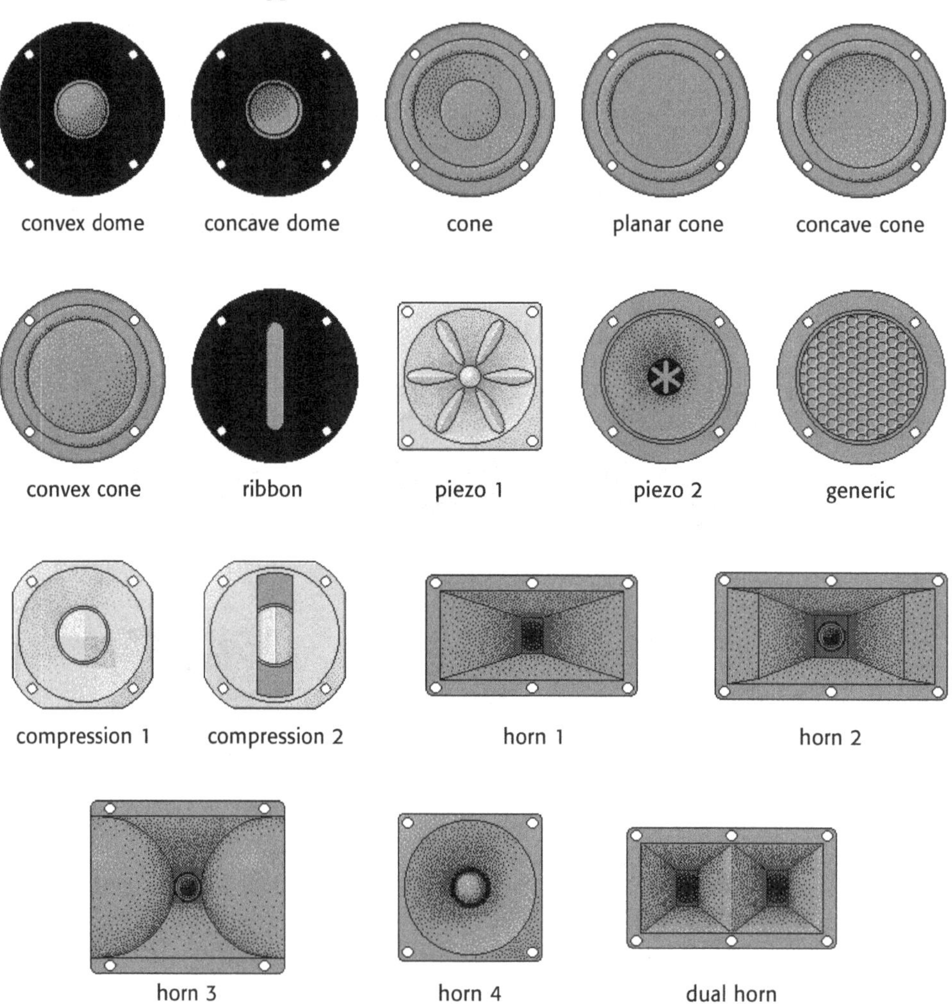

Dimensions The various mounting dimensions of the driver. Units: mm, cm, m or in. Refer to Appendix C (open back drivers) and Appendix D (sealed back drivers) for an example of the dimensions for each driver shape. Before entering a dimension, remember to select the desired units by clicking on its unit label.

Square Occasionally a driver will have a square rather than round magnet. Check the "Square" checkbox when this is so. When it is checked, dimension "D" will be assumed to be the length of one side of the magnet rather than the diameter.

M-Vd The mounting volume of the driver. Units: liters, cu.cm, cu.m, cu.ft or cu.in. This is the volume displaced by the driver in the box. If enough dimensions have been entered, X•over Pro can estimate this volume. When this is possible, an "Est" button will appear beside the "M-Vd" label.

5 Using the "Enter Response" tab enter the acoustical response of the driver. This step is optional and requires the measured acoustic response data to be normalized to the response curve predicted by the driver's Thiele-Small parameters. See Chapter 2 (pages 113-119) for a description of the normalization process. **Important:** Before acoustic data can be normalized to the predicted T-S response, the driver's parameters must be entered in Step 3 above and then the driver must be "loaded" into the design. If the driver has an open back and was measured in a test box, the box information must also be entered into the design (with the "Impedance EQ" tab of the Driver Properties window) in order to duplicate the conditions used for the acoustical measurement. This will enable the program to generate the predicted response curve.

6 Add the new driver to the driver database by clicking on the "Save" button. Before the new driver is added to the database, X•over Pro will check the company name to make sure that it is a valid company in the database and it will check its model name to make sure that it is not a duplicate driver for the company. Then it will check to see if the minimum parameters have been entered. If the data checks out, the driver will be added to the end of the database.

Changing an Existing Driver

Changing an existing driver is easy. First, use the Driver Locator window to find it. Then use the "Description", "Parameters", "Dimensions" and "Response" tabs of the Edit Database Driver Data window to make the desired changes. Then click on the "Save" button.

Deleting an Existing Driver

First, use the Driver Locator window to find the unfortunate driver. Then, with firm resolution, click on the "Delete the Selected Driver" button. X•over Pro will confirm your request before executing it.

Editing Companies

The Edit Companies window is used to add a new company or edit or delete an existing one. To open the Edit Companies window, click on the "Edit" button to the right of the "Companies" drop-down list on the "Description" tab of the Edit Database Driver Data window. You can also edit the companies directly from the main window by selecting "Database > Edit Company Data" from the Edit menu as shown below:

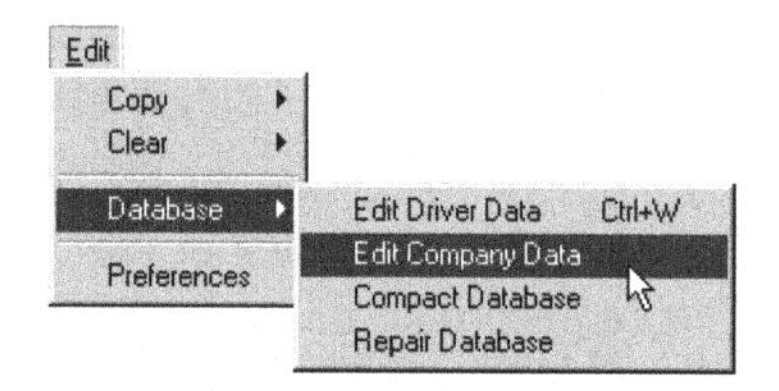

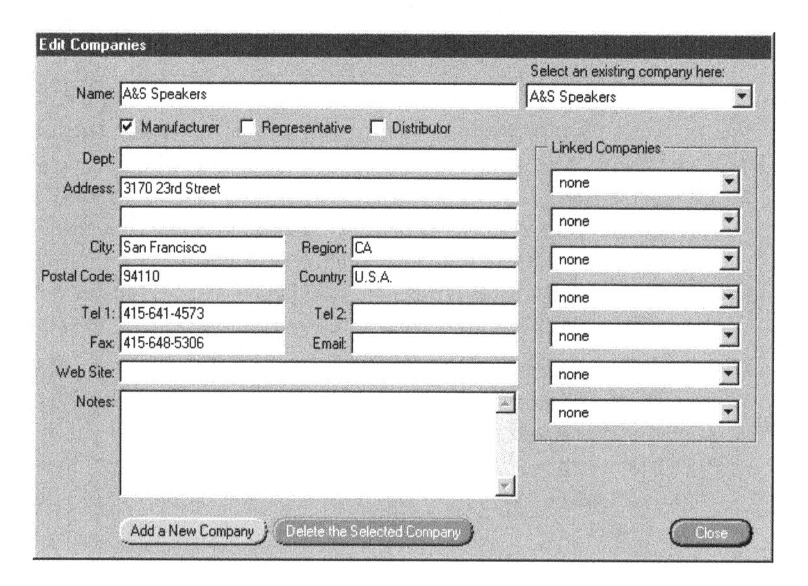

Adding a New Company

To add a new company, click on the "Add a New Company" button and enter its name and other information. A "Save" button will appear when you begin to enter the company. After you have finished, click on the "Save" button to enter the company into the database. *Note: Two or more companies cannot have the same name.*

Important: Drivers should only be linked to companies that are listed as "manufacturers". Otherwise they may not appear in the database. The "representative" and "distributor" settings are discussed on the next two pages.

9

Editing an Existing Company
Begin by selecting the company with the "Select an existing company here" drop-down list as shown in the sample below:

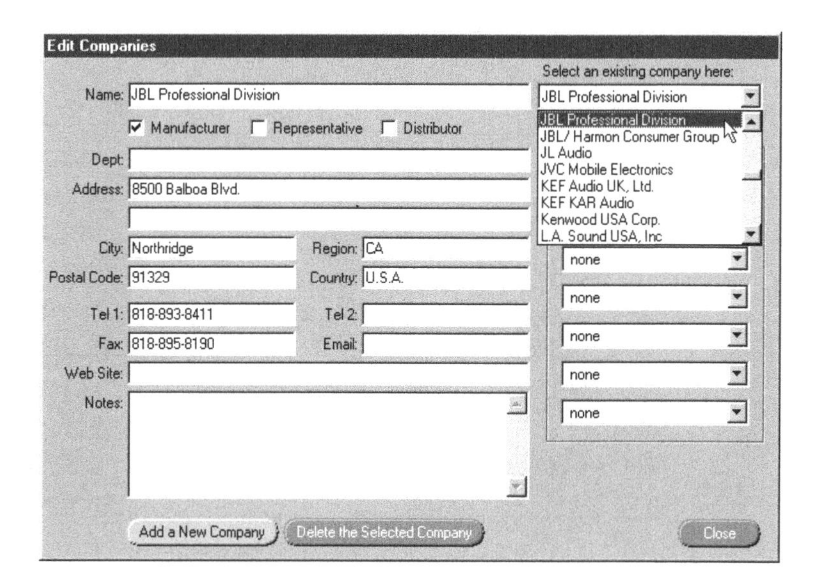

Make the desired changes. The "Save" button will appear when the first change is made. After you have finished, click on the "Save" button.

Deleting an Existing Company
Select the unfortunate company with the "Select an existing company here" drop-down list as shown above. Then click on the "Delete the Selected Company" button. X•over Pro will confirm your request before executing it.

Linking Representatives & Distributors
If the manufacturer of a driver does not allow direct communication, you may need to enter the company information for one or more representatives and/or distributors. A representative or distributor is entered just like a manufacturer but the "Representative" and/or "Distributor" checkboxes are checked instead of the "Manufacturer" checkbox.

9

After the representatives and/or distributors have been entered, return to the company that is the manufacturer. Then use the drop-down lists under "Linked Companies" to select the representatives and/or distributors.

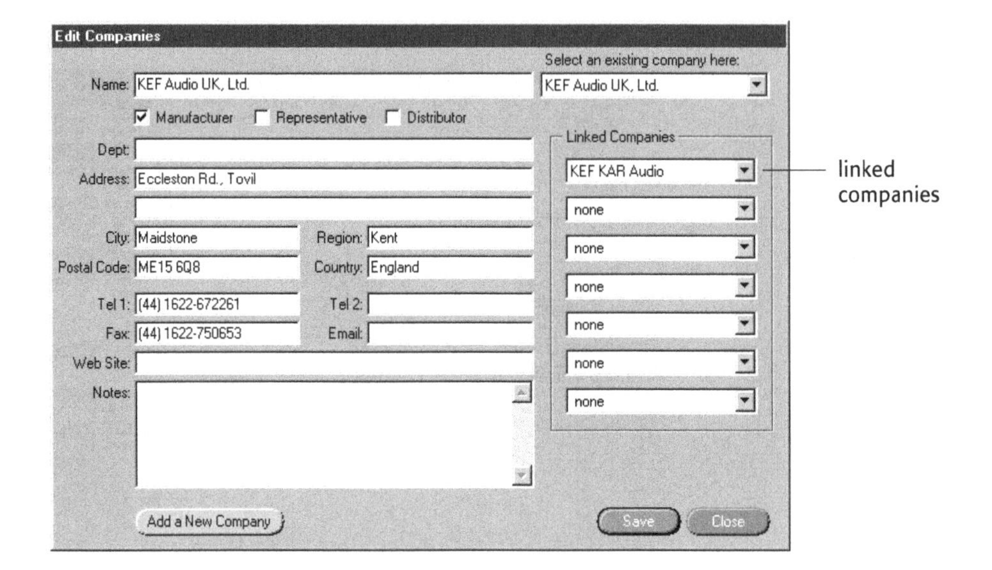

Up to seven representative or distributor companies can be linked to a manufacturer. The linked companies will always appear after the manufacturer in the information box of the Driver Locator.

Important: Do not link representatives and distributors to each other. They should only be linked to a manufacturer.

Note: It is okay for a manufacturer to also be a representative and/or distributor. This situation, although rare, occurs when a company both manufactures its own line of drivers and at the same time distributes another company's line of drivers. In such cases, it is okay to link one manufacturer to another manufacturer.

Compacting the Database

Deleting drivers and/or companies from the database does not decrease its size because that would require the entire database to be resaved, greatly slowing its operation. To compact the database after one or more drivers or companies have been deleted from it, select the "Database > Compact Database" command from the Edit menu of the main window.

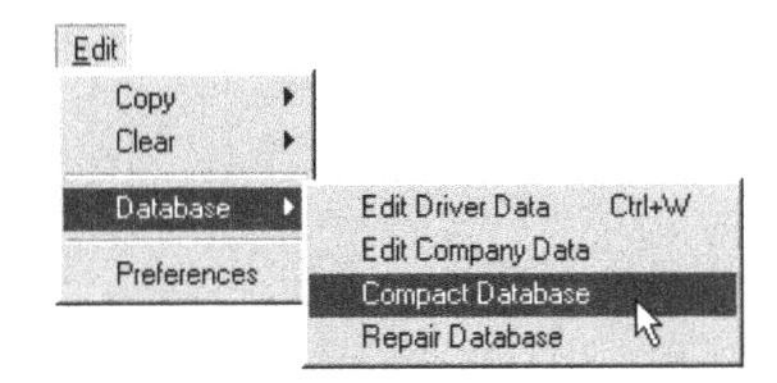

Note: A backup copy of the uncompacted database will be made before the database is compacted. Its name is "htaudio.bak". To restore it, delete "htaudio.mdb" and rename the backup to "htaudio.mdb". The database should be in the same location as the X•over Pro program. This is "C:\Program Files\HT Audio" if the default settings were used when X•over Pro was installed.

Repairing the Database

If your computer looses power or "crashes" while the database is being edited, the database may become "corrupt". If, when you attempt to open the database, you ever receive an error message saying that the database is unrecognizable or is not a Microsoft Access database, you may be able to repair it with the "Database > Repair Database" command in the Edit menu of the main window.

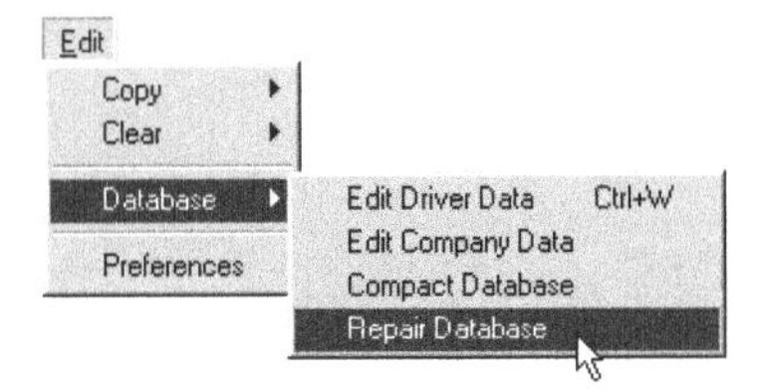

Important: Some data in corrupted portions of the database may be lost when the database is repaired. If this happens, you can reinstall a fresh copy of the database with the X•over Pro installation CD. However, any changes that you made will be lost.

Note: A backup copy of the unrepaired database will be made before the repair is attempted. Its name is "htaudio.bak". To restore it, delete "htaudio.mdb" and rename the backup to "htaudio.mdb". The database should be in the same location as the X•over Pro program. This is "C:\Program Files\HT Audio" if the default settings were used when X•over Pro was installed.

10 Tools

X•over Pro provides access to several helpful tools in its Tools menu as shown below. This chapter will describe each one.

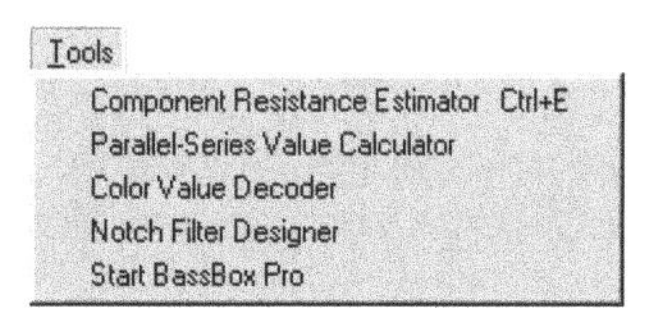

Component Resistance Estimator

The Filter Component Resistance Estimator (shown below) will estimate the equivalent series resistance (ESR) of many popular types of capacitors and the DC resistance (DCR) of many types of inductors. To open it, select the "Component Resistance Estimator" command from the Tools menu (keyboard shortcut Ctrl+E) or right-click () on an ESR or DCR input box in the main window.

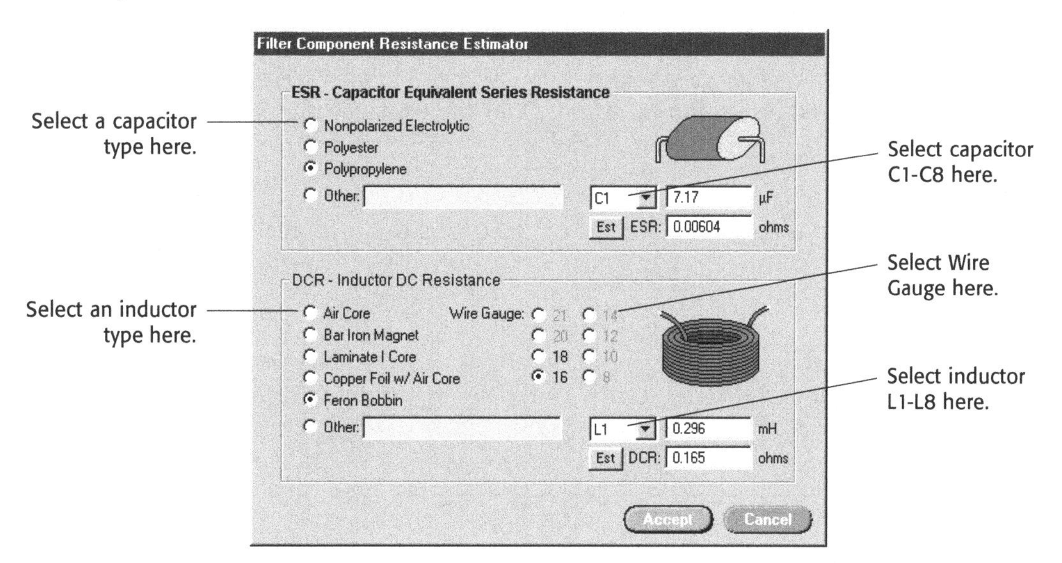

Follow these steps to estimate capacitor ESR:

1. Select an existing capacitor (C1, C2, C3, C4, C5, C6, C7 or C8) with the drop-down list provided or enter a capacitance value in the capacitance input field to the right of it.
2. Select a known capacitor type (Nonpolarized Electrolytic, Polyester or Polypropylene).
3. Click the "Est" button and the ESR will be estimated.

Follow these steps to estimate inductor DCR:

1 Select an existing inductor (L1, L2, L3, L4, L5, L6, L7 or L8) with the drop-down list provided or enter an inductance value in the inductance input field to the right of it.
2 Select a known inductor type (Air Core, Bar Iron Magnet, Laminate I Core, Copper Foil w/ Air Core or Feron Bobbin).
3 Select an available wire gauge. Which wire gauges are available depends on the inductor type. Available wire gauges use black lettering. Unavailable wire gauges use grey lettering. A summary is listed below:

   Air Core: .... 21, 20 18, 16 and 14 gauge
   Bar Iron Magnet: .... 18 and 16 gauge
   Laminate I Core: .... 18 gauge
   Copper Foil w/ Air Core: ...... 16, 14 and 12 gauge
   Feron Bobbin: .... 18 and 16 gauge

4 Click the "Est" button and the DCR will be estimated.

Finally, the "Other" capacitor and inductor type is provided so additional component types can be included in the printouts. When the "Other" type is chosen or when an unavailable inductor wire gauge is selected, you will need to manually enter the ESR or DCR. If no value is entered, X•over Pro will assume that the resistance is very low.

Parallel-Series Value Calculator

When designing a crossover network it is common to need an easy way of calculating the net value when two or more components are combined. Combining components is often necessary to get a desired value.

This tool will allow you to add up to four components of the same type such as four resistors or four capacitors. They can each have a different value and they can be added in series or parallel (both are shown below).

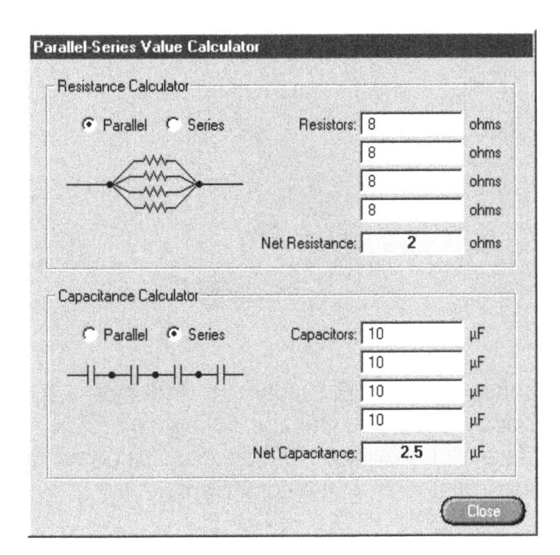

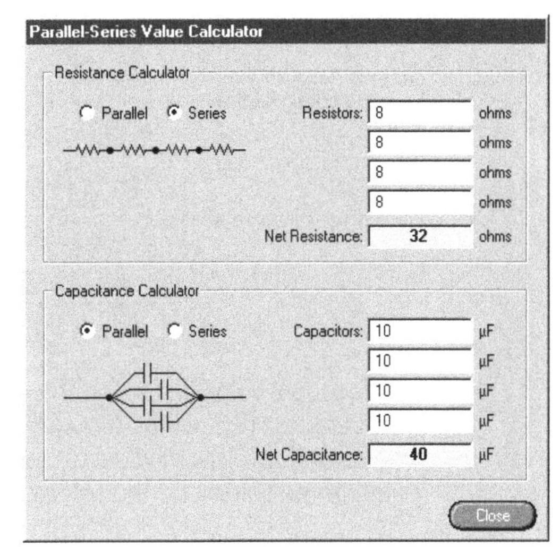

Resistors and Capacitors add in opposite ways. When resistors are wired end-to-end in series, their values add as a simple sum. But capacitors add as a simple sum when they are wired side-by-side in parallel. Inductors add like resistors but they are not included because they are prone to magnetic coupling and this can affect their value. For this reason multiple inductors are seldom added together and great care must be taken with their location and mounting. (For mounting information see Chapter 6 of the *Crossover Network Designer's Guide* earlier in the manual.)

Color Value Decoder

The value of many capacitors, inductors and resistors is identified by decoding color bands or stripes that are painted on them. The Color Value Decoder window will help you decode these values.

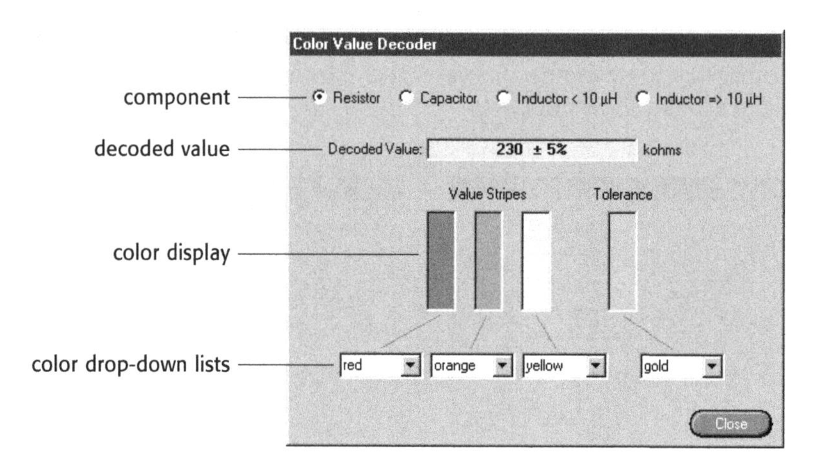

To decode a color, first select the component ("Resistor", "Capacitor", "Inductor < 10 µH" or "Inductor => µH") at the top of the window. Next select the appropriate color from each of the drop-down lists provided at the bottom of the window. The decoded value will be calculated immediately and the selected colors will appear in the color display.

10

Notch Filter Designer

Ever wanted to design a simple notch filter and nothing more? If so, this tool is for you! The Notch Filter Design window is a stand-alone tool that can help you design series and parallel notch filters. The filters it designs are not displayed in the graphs or linked to any particular driver in X•over Pro.

Each of the notch filters uses a single capacitor, inductor and resistor. **It is important to note that the resistor will need to be adjusted after an inductor is selected.** The adjustment is to subtract the DCR of the inductor from the calculated value of the resistor. The final resistor value will therefore be a little smaller.

Also note that lower filter frequencies require larger inductors. Below 200 Hz, the inductor can become quite large and therefore have considerable DCR. For this reason, these filters are best used at frequencies above 200 Hz.

Series Notch Filter

In a series notch filter, the capacitor, inductor and resistor are wired in series with each other. Two methods (A & B) are provided for designing a series notch filter. Method A (shown below) is best suited for equalizing the resonance peak of a driver because it uses the following driver parameters: free air resonance (Fs), voice coil DC resistance (Re) and the electrical and mechanical resonance magnification (Qes and Qms).

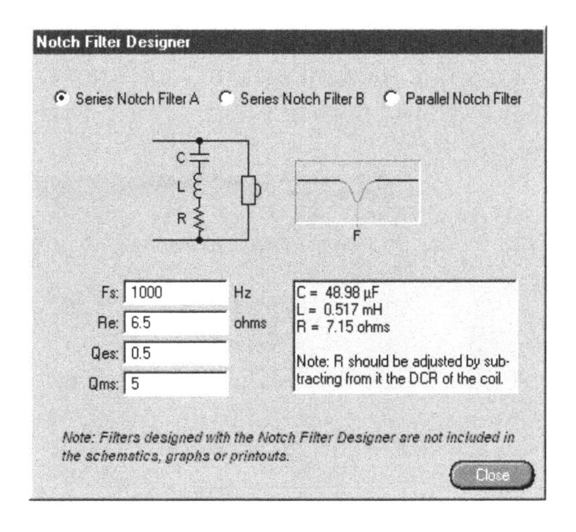

Method B (shown below) is more general and requires only two parameters: frequency (F) and impedance (Z).

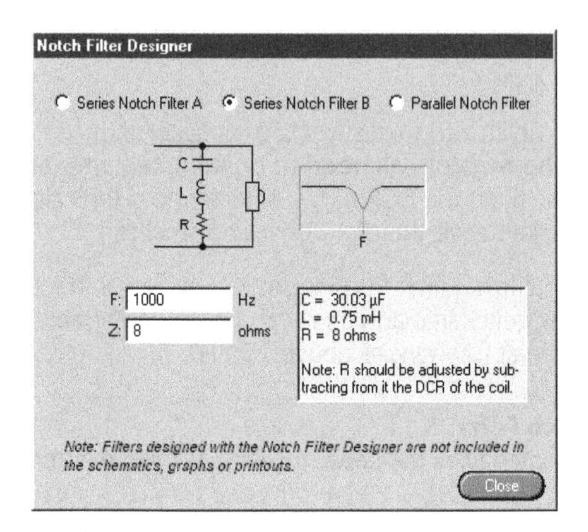

Parallel Notch Filter

In a parallel notch filter (shown below), the capacitor, inductor and resistor are wired in parallel with each other. It has the advantage of working well when a broad, shallow notch is desired. It allows you to specify the width of the notch with parameters F_1 and F_2. The center frequency is parameter F.

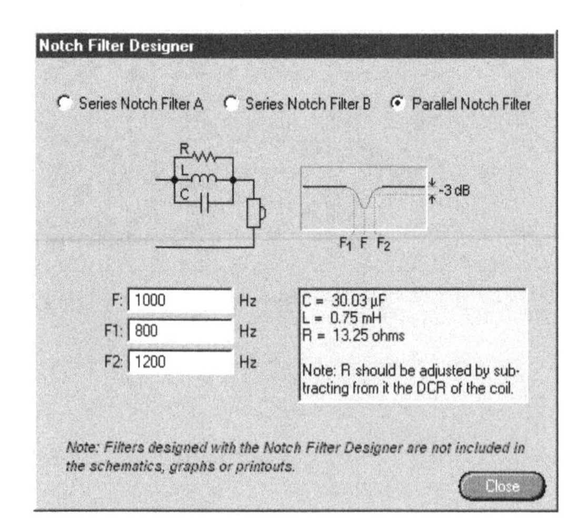

BassBox Pro

The optional "Start BassBox Pro" command of the Tools menu launches the BassBox 6 Pro program, if available. This command is enabled if you have purchased a BassBox 6 Pro license and the program is installed on your computer.

BassBox 6 Pro (shown below) is a speaker box design program from Harris Tech. Using it you can design a box, calculate box and vent dimensions, create a parts list ("cut sheet") with cut angles. It provides separate acoustic modeling capabilities for both the driver and listening environment and it provides a total of 9 graphs to analyze the performance of the speaker including the acoustic power, electric input power, cone displacement and vent air velocity. It also includes test procedures for measuring the Thiele-Small parameters of open back drivers and passive radiators (requires basic test equipment).

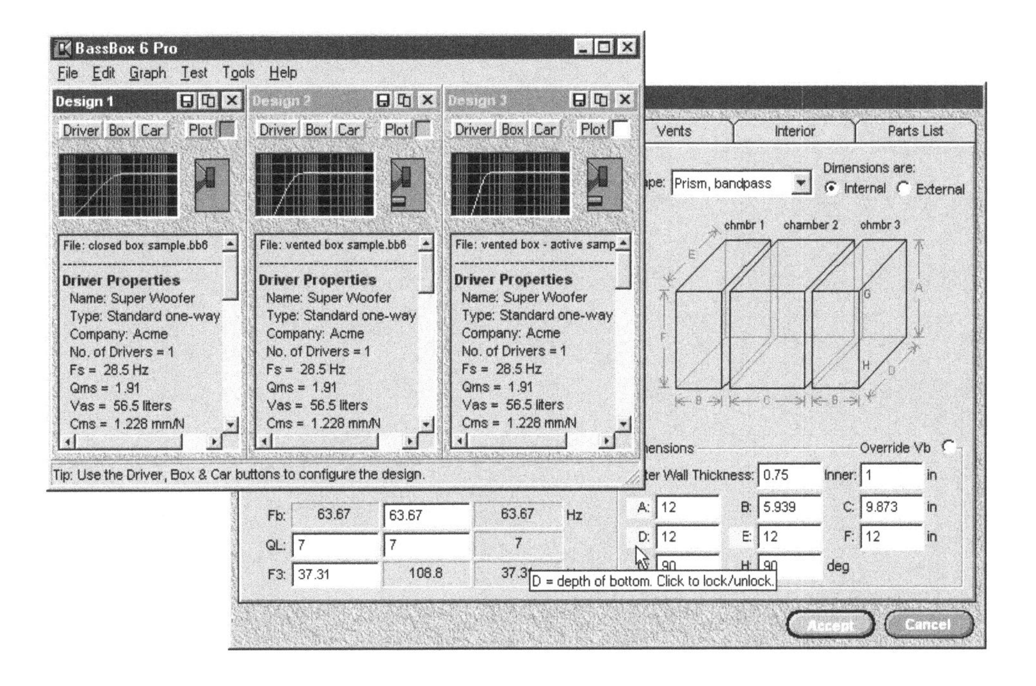

BassBox 6 Pro can import filter information from a crossover network from X•over Pro, making it easy to observe its effect while designing the box. Together the two programs provide a comprehensive speaker design system. Design the box and its dimensions with BassBox 6 Pro. Design the crossover network and create a schematic with X•over Pro. And with its ability to import information from BassBox 6 Pro, X•over Pro can model the full speaker system with the net response plot line in its Normalized Amplitude Response and System Impedance graphs.

If the "Start BassBox Pro" command is not enabled in the Tools menu and yet the program is properly installed on your computer, then its path entry in the Preferences window is probably blank. This is easy to correct by manually entering the location of BassBox 6 Pro on the "General" tab of the Preferences window. See Chapter 11 (pages 211-212) for details.

To obtain information about BassBox 6 Pro and how to purchase a software license for it, please visit Harris Tech's website at www.ht–audio.com or contact them directly via email at sales@ht–audio.com.

11 Preferences

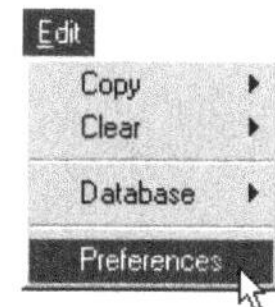

X•over Pro offers an extensive set of default parameters to tailor the program to your preferences. These "preference" settings are available in the Preference window. To open it, select "Preferences" from the Edit menu of the X•over Pro main window as shown at right:

The Preferences window is divided into 7 tabs which will be discussed in detail next. When the window first opens, it always opens to the first "General" tab.

General

The "General" tab contains mostly global settings that affect multiple areas of X•over Pro.

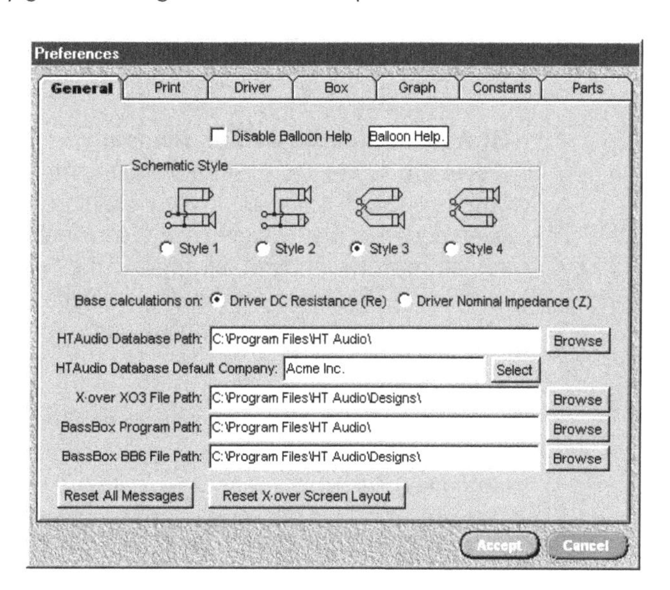

Balloon Help

Extensive balloon help is provided in X•over Pro. It can be turned off with the "Disable Balloon Help" checkbox.

Schematic Style

Four choices are available for the schematic or circuit diagram that is displayed and printed by X•over Pro. The default style is selected here. The style can also be temporarily changed by double-clicking (🖱🖱) on the schematic in the main window of the program.

Base calculations on

Whenever the component values of a crossover network filter or impedance equalization network are calculated, the load impedance must be known. X•over Pro offers you the choice for this load impedance. It can use the DC resistance (Re) of the driver or the nominal impedance (Z) of the driver. The setting you choose will be the basis for all crossover network filter and all impedance equalization network calculations. The setting will also be the default for L-pad calculations (although the L-Pad Attenuator window will allow you to set the total load impedance separately for each L-pad).

Note: Harris Tech recommends using Re for most calculations. They say it often produces the best results.

Whenever the selected load value is unknown, X•over Pro will attempt to estimate it from the known value. To accomplish this, the program assumes that Z is 1.2 times larger than Re. For example, if Re is selected and only Z is known, the program will estimate Re by dividing 1.2 into Z. Conversely, if Z is selected and only Re is known, the program will estimate Z by multiplying Re by 1.2. This is done as needed for those drivers that lack the desired value.

HT Audio Database Path
A "path" is like a street address. It tells your computer where to go to find something. The HT Audio Database Path is the location of the driver database file on your computer's hard drive and this path setting was set during the installation of X•over Pro. You shouldn't normally need to change it. If the default settings were used during installation, the path will be "C:\Program Files\HT Audio\". The database file name is "htaudio.mdb". If you ever move it, or want to keep multiple copies in different folders, you can use this setting to tell X•over Pro where to find it. A "Browse" button is provided to help you select a new path, if desired.

HT Audio Database Default Company
X•over Pro allows you to set a default company for the driver database. Each time the database is opened, it will open with the default company selected and its drivers will have been added to the "Drivers Found" list. There are two ways to enter a company name. It can be manually entered directly into the "HT Audio Database Default Company" input box or it can be selected from the driver database using the "Select" button. The latter method is preferred because the company name must be spelled exactly as it is in the database.

X•over XO3 File Path
When you save a crossover network or filter design, X•over Pro creates a file with the extension "xo3". This path setting tells X•over Pro where to find these files. If the default settings were used during installation, the path will be "C:\Program Files\HT Audio\Designs\". A "Browse" button is provided to help you select a new path, if desired.

BassBox Program Path
BassBox Pro is a speaker box design program from Harris Technologies. It can be launched from X•over Pro and this path setting tells X•over Pro where to find it. If the default settings were used during installation, the path will be "C:\Program Files\HT Audio\". A "Browse" button is provided to help you select a new path, if desired.

BassBox BB6 File Path
Driver and box parameters can be imported into an X•over Pro crossover network design from BassBox Pro or BassBox Lite. This enables X•over Pro to show the net response of the speaker system, including the crossover network and box design. BassBox Pro/Lite design files end with the extension "bb6" and this path setting tells X•over Pro where to find them. If the default settings were used during installation, the path will be "C:\Program Files\HT

Audio\Designs\". A "Browse" button is provided to help you select a new path, if desired.

Reset All Messages
Some of X•over Pro's message windows can be disabled by clicking on a "Don't show this message in the future" checkbox in the message window. Clicking on the "Reset All Messages" button will uncheck all these settings so that all message windows will appear again.

Reset X•over Screen Layout
X•over Pro remembers the last location of many of its various windows so that they will appear in their previous locations on the screen each time that they are opened. What would happen if the windows were positioned around the screen while it was at a very high resolution, such as 1600 x 1200 pixels, and then the screen was later reduced to a much smaller resolution, such as 640 x 480 pixels? In such cases, some of X•over Pro's windows may no longer be visible on the screen. Clicking the "Reset X•over Screen Layout" button will restore all X•over Pro windows to their default positions so they will be visible.

Print

Most of the print options have a "preference" setting so that a default printout can be created. These settings are available in the "Print" tab of the Preferences window.

Enable Color Graphics
This option enables the graphs to print with color plot lines. This option will have no effect on the logo picture. If a color logo picture is selected and a color printer is used, the logo will still print in color even if the "Enable Color Graphics" option is turned off.

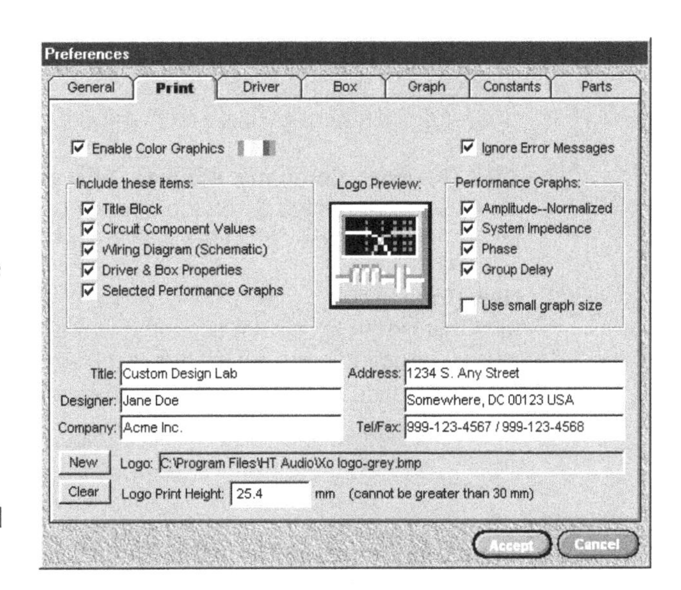

Ignore Error Messages
Normally this option should be enabled. If you are having print problems, it can be unchecked to enable error messages to aid with troubleshooting.

Include These Items
There are five items that can be included in a printout. Check the ones that you want to include. They are summarized on the next page.

11

Title Block Includes the Title line, Designer line, Company line, Address lines, Tel/Fax line, Notes lines and the logo picture. It is printed on page one.

Circuit Component Values Includes a list of all the capacitors, inductors and resistors in the crossover network or filters, impedance equalization networks and L-pads. It also includes general information about the crossover network filters such as the filter frequencies and a list of the filter types used.

Wiring Diagram (Schematic) Includes a circuit diagram of the crossover network or filters, impedance equalization networks and L-pads. It also includes configuration wiring details of the drivers.

Driver & Box Properties Includes a list of the driver and box parameters.

Selected Performance Graphs Includes the graphs which are checked in the "Performance Graphs" section.

Logo Preview

The logo preview shows the picture that is presently selected for the logo. The logo prints in the title block of the first page. A smaller version also prints in the header of successive pages. Use the "New" button to select a different picture for the logo. Use the "Clear" button to clear the logo so none will print.

Performance Graphs

Select the graphs that you want to print when the "Selected Performance Graphs" checkbox is checked in the "Include these items" section. The "Use small graph size" option causes a smaller graph size to be used so that more graphs will fit on a page.

Title, Designer, Company, Address, Tel/Fax

These lines are printed at the top of page one in the title block when the "Title Block" option is enabled under "Include these items" section.

New (logo)

Click this button to select a different picture for the logo. Several picture file formats are supported including: bmp, dib, gif, jpg, wmf, emf, ico and cur. Three versions of the X•over Pro icon are included with the program. They are:

Xo logo-color.bmp	a color version of the X•over Pro icon
Xo logo-grey.bmp	a greyscale version of the X•over Pro icon
Xo logo-mono.bmp	a black & white version of the X•over Pro icon

These pictures are located in the same folder as X•over Pro ("C:\Program Files\HT Audio").

Clear (logo)

Clears the logo picture so that no logo will print.

Logo Print Height
Sets the height of the logo. The maximum allowable height is 1.2 inches or 30 mm. The logo picture will be proportionally scaled to this height when it is printed.

Driver

There are several "preference" settings for drivers. These are depicted below in the "Driver" tab of the Preferences window.

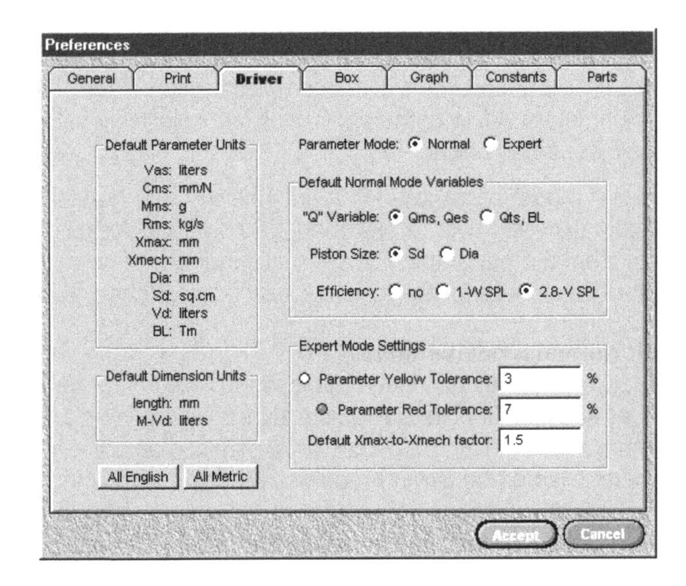

Default Parameter Units
Sets the default units for the driver parameters throughout the program. This affects the Driver Properties window, Edit Database Driver Data window, Driver Locator and printouts.

To change the units, click once (🖱) on the desired units label. Each time you click, it will advance to the next available setting. *Note: The default values here can often be overridden in the program by clicking on the individual units labels in the various windows.*

Default Dimension Units
Sets the default units for the driver dimensions throughout the program. This affects the Driver Properties window, Edit Database Driver Data window and the printouts.

To change the units, click once (🖱) on the desired units label. Each time you click, it will advance to the next available setting. *Note: The default values here can often be overridden in the program by clicking on the individual units labels in the various windows.*

11

All English
Click this button to quickly change all the default parameter and dimension units to English.

All Metric
Click this button to quickly change all the default parameter and dimension units to metric.

Parameter Mode
There are two operating modes for the "Parameters" tab of the Driver Properties window. These modes are "normal" and "expert". *Note: The "Parameters" tab in the Edit Database Driver Data window is always in "expert mode" to provide status indicators.*

The "normal mode" presents a less complex version of the "Parameters" tab with fewer driver parameters. A few of the parameters are selectable with the "Default Normal Mode Variables" described below.

The "expert mode" presents a full set of Thiele-Small and electromechanical parameters in the "Parameters" tab. The "expert mode" automatically tests many of the parameters to see if any of them are out of tolerance. A status indicator shows whether they passed or failed. The tolerance settings for the tests are selectable with the "Expert Mode Settings".

Default Normal Mode Variables
Three sets of parameters are selectable in the "normal mode". One set selects between Qms/Qes and Qts/BL. The second set selects between Sd and Dia. The third set selects between ηo, 1-W SPL and 2.8-V SPL. The selected parameters will be displayed when the "Parameters" tab of the Driver Properties window is first selected. *Note: The user can override the default values by clicking (🖱) on the appropriate parameter labels in the Driver Properties window.* Select the defaults which best fit the driver data you anticipate using.

Expert Mode settings
The "expert mode" settings configure the automatic self-analyzing feature.

Parameter Yellow Tolerance Sets the ± percent variation that is allowed before a status indicator is turned yellow.

Parameter Red Tolerance Sets the ± percent variation that is allowed before a status indicator is turned red.

Default Xmax-to-Xmech factor Is multiplied times Xmax to estimate Xmech and visa versa. Xmax is the maximum linear excursion measured from rest in one direction. Xmech is the maximum mechanical excursion measured from rest in one direction.

Box

There are several "preference" settings for boxes. These are depicted below in the "Box" tab of the Preferences window.

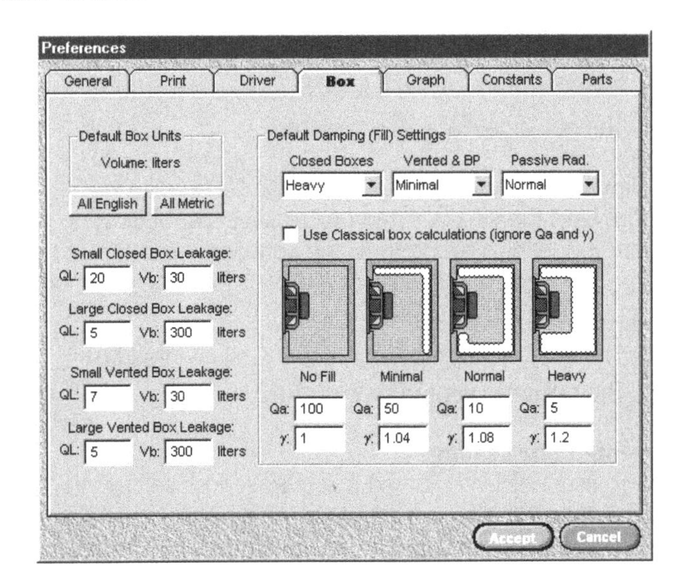

Default Box Units

Sets the default units for the box volume. This affects the "Impedance EQ" tab of the Driver Properties window and the printouts.

To change the units, click once (🖱) on the desired units label. Each time you click, it will advance to the next available setting. *Note: The default values here can often be overridden in the program by clicking on the individual unit labels in the various windows.*

All English

Click this button to quickly change the default box units to English.

All Metric

Click this button to quickly change the default box units to metric.

Small Closed Box Leakage

X•over Pro uses the small closed box leakage loss settings to estimate the value of QL when necessary. It does this by interpolating between the small box and large box values to estimate a value for the net volume of a closed box. The small box value represents the highest value of QL and a typical value is 20.

QL The Q of the leakage losses of a small closed box.

Vb The net small closed box volume.

Large Closed Box Leakage
X•over Pro uses the large closed box leakage loss settings to estimate the value of QL when necessary. It does this by interpolating between the small box and large box values to estimate a value for the net volume of a closed box. The large box value represents the lowest value of QL and a typical value is 5.

QL The Q of the leakage losses of a large closed box.

Vb The net large closed box volume.

Small Vented Box Leakage
X•over Pro uses the small vented box leakage loss settings to estimate the value of QL when necessary. It does this by interpolating between the small box and large box values to estimate a value for the net volume of a vented box. The small box value represents the highest value of QL and a typical value is 7.

QL The Q of the leakage losses of a small vented box.

Vb The net small vented box volume.

Large Vented Box Leakage
X•over Pro uses the large vented box leakage loss settings to estimate the value of QL when necessary. It does this by interpolating between the small box and large box values to estimate a value for the net volume of a vented box. The large box value represents the lowest value of QL and a typical value is 5.

QL The Q of the leakage losses of a large vented box.

Vb The net large vented box volume.

Default Damping (Fill) Settings
The "Default Damping Settings" control the way X•over Pro models the addition of acoustic absorption material to the interior of the box.

None, Minimal, Normal, Heavy Selects the amount of damping material that will be automatically entered for the new box of an open-back driver. There are separate settings for the different box types. *Note: The vented and bandpass boxes share the same setting because they both have vents.* If desired, the user can override these default settings later with "Impedance EQ" tab of the Driver Properties window.

Use Classical box calculations (ignore Qa and γ) When enabled, X•over Pro will ignore the damping settings and use "classical" box calculations. This means that Qa and γ will be ignored in the box calculations. Normally, this control should be turned off so that X•over Pro is allowed to adjust the box calculations to compensate for acoustic "fill".

Qa Sets the values of the Q of the absorption losses (Qa) for the different damping settings. The factory settings for X•over Pro are 100, 50, 10 and 5 for None, Minimal, Normal and Heavy damping settings, respectively.

γ (gamma) Sets the γ values for the different damping settings. Gamma is the ratio of heat at constant pressure to that at constant temperature for the air in the box. X•over Pro uses a delta (ratio of change) value that represents the difference from normal air (which is around 1.4). The factory settings for X•over Pro are 1.00, 1.04, 1.08 and 1.20 for None, Minimal, Normal and Heavy damping settings, respectively.

Graph

There are many "preference" settings for graphs. These are depicted below in the "Graph" tab of the Preferences window.

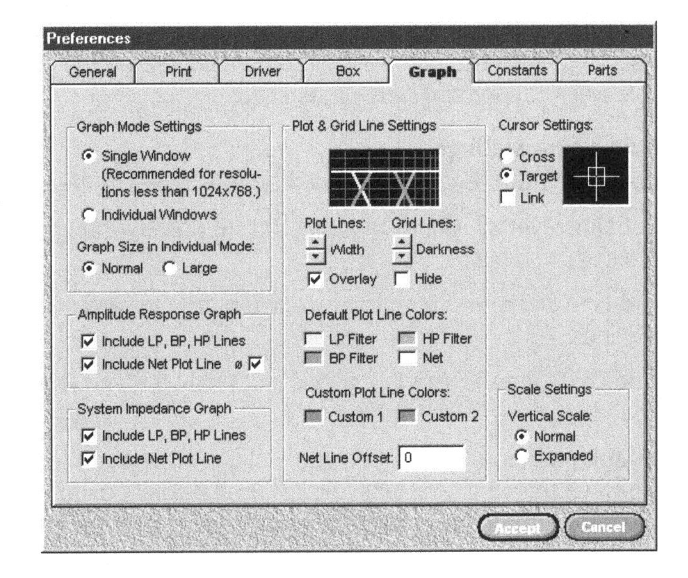

Graph Mode Settings

Mode Selects the default graph mode. Two graph modes are available. The "Single Window" Mode displays all graphs one at a time in a single window. This is recommended for computers with video resolutions less than 1024 x 768 pixels (XGA). The "Individual Windows" mode displays each graph in its own individual window, allowing multiple graphs to be viewed simultaneously.

Graph Size in Individual Mode Sets the default size of the graph windows in the "individual windows" mode.

Amplitude Response Graph

Selects the plot lines which will be included in the Normalized Amplitude Response graph. The "Include LP, BP, HP Lines" option causes the individual response of each filter to be plotted with a separate line. The "Include Net Plot Line" option causes a composite plot line to be included. *Note: Only the individual LP, BP, HP plot lines are available when "Separate Filters" is selected.*

There are two ways to calculate the net response. When "ø" (phase) is turned on, it is the sum of both the magnitude and the phase of the volume velocities of the filters. When "ø" is turned off, it is the sum of the magnitudes only and will not show phase cancellations.

System Impedance Graph

Selects the plot lines which will be included in the System Impedance graph. The "Include LP, BP, HP Lines" option causes the individual response of each filter to be plotted with a separate line. The "Include Net Plot Line" option causes a composite plot line to be included. The composite or "net" plot line is created by adding the parallel impedance of the low-pass, band-pass and high-pass filters. *Note: Only the individual LP, BP, HP plot lines are available when "Separate Filters" is selected.*

Plot & Grid Line Settings

Plot Line Width Use the up/down buttons to select the width of the plot lines.

Plot Line Overlay Check this checkbox to prevent the graphs from being erased before each new plot.

Grid Line Darkness Use the up/down button to select the desired darkness of the grid lines.

Hide Grid Lines Check this checkbox to hide the grid lines. The graphs will then be plotted on a solid black background.

Default Plot Line Colors Sets the plot line color given to each filter and the net plot line when the program starts. Click (🖱) on a color box to change the color. Each click will cause the color to advance through the 12 possible colors, including the two custom colors.

Custom Plot Line Colors Enables you to set two custom colors to be used for any of the plot lines. Click (🖱) on one of the two custom color boxes and the standard Windows Color window shown at right will open so you can select a custom color.

Note: The available color range and color quality will depend upon the capabilities and settings of the computer's video system.

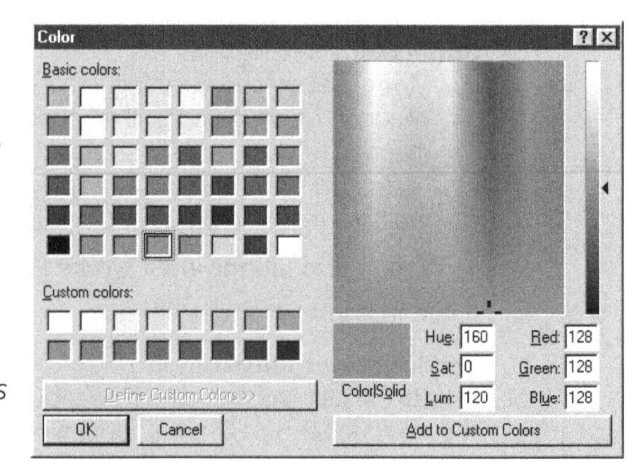

Net Line Offset If desired, the net plot line can be offset above or below the individual plot lines. Use a positive value to move the net plot line above the LP, BP, HP filter plot lines and visa versa. Three graphs, the mini preview graph, Normalized Amplitude Response graph and System Impedance response graph, include a net plot line. The "Net Line Offset" is interpreted as follows for each graph:

Normalized Amplitude Response graph: Offset = dB
System Impedance Response graph: Offset Value x 2 = ohms

Cursor Settings
The shape of the cursor can be selected with the "Cross" and "Target" options. The cursors in all graphs can be linked with the "Link" option. When linked, all cursors will move in unison, making it easy to control the cursors in multiple individual graphs.

Scale Settings
Two vertical graph scale options are available and this setting selects the default. The range of the "normal" and "expanded" scales varies depending on the graph.

Constants

There are two "constants" used by X•over Pro which pervade the calculations of the program. These are the density of air and the velocity of sound in air. They can be changed using the "Constants" tab of the Preferences window.

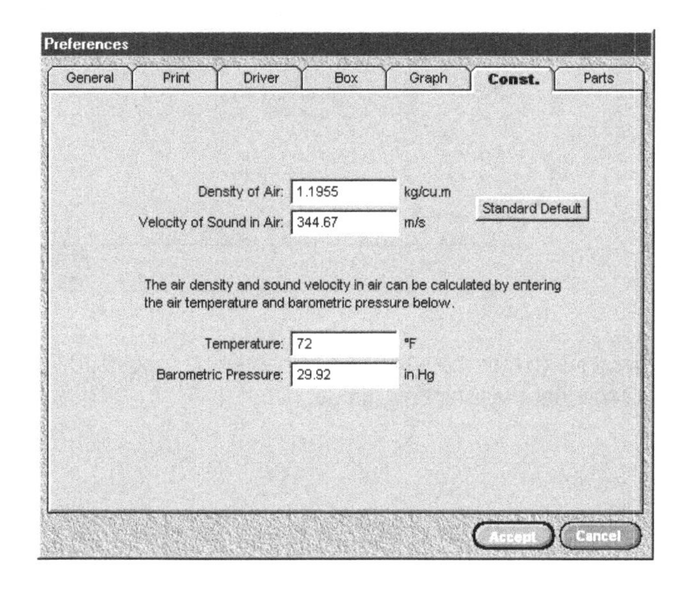

There are two ways to change the values of these constants. A new air density or sound velocity value can be manually entered into its respective input box. Or, the temperature and barometric pressure can be entered and the air density and the sound velocity will be automatically calculated from them. Before making changes, remember to click on the unit labels to select the desired units.

To restore the factory values for the constants, click on the "Standard Default" button.

Parts

When X•over Pro calculates the values of the capacitors and inductors in a crossover network or filter it will also estimate the resistance of the components. This is the ESR or equivalent series resistance of capacitors and the DCR or DC resistance of inductors. In order to do this, X•over Pro needs to know what kind of capacitor or inductor is used. The "Parts" tab allows you to select a default capacitor and inductor type for this purpose.

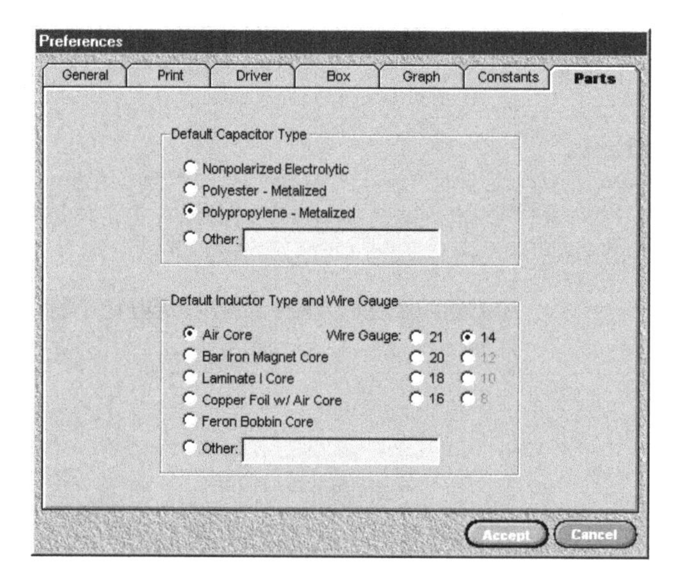

See Chapter 3 (pages 139-140) and Chapter 10 (pages 203-204) for more information about component resistance estimation.

Appendix A: Command Shortcuts

The following keyboard shortcuts can be used to swiftly access many commonly-used X•over Pro commands. A "+" (plus) sign means to hold down the first key(s) while pressing the last key. For example, Ctrl+P means to press and hold the Ctrl key and then press and release the P key. Release the Ctrl key after the P key has been released.

General

F1	Open the on-screen manual.
Ctrl+P	Print the crossover network / filter design.
Ctrl+Q	Close X•over Pro.

Files

Ctrl+O	Open an existing crossover network design from disk.
Ctrl+S	Save design changes to disk in an X•over Pro XO3 design file.

Design

Ctrl+N	Begin a new crossover network / filter design.
Ctrl+D	Open the Driver Properties window.

Network

F2	Configure X•over Pro to design a 2-way crossover network.
F3	Configure X•over Pro to design a 3-way crossover network.
F4	Configure X•over Pro to design separate filters.
Ctrl+R	Recalculate the network / filter component values.
Ctrl+L	Link or unlink the adjacent filter frequencies.
Ctrl+A	Open the L-Pad Attenuator window.
Ctrl+E	Open the Filter Component Resistance Estimator window.

Graphs

Ctrl+F1	Open or select the Normalized Amplitude Response graph.
Ctrl+F2	Open or select the System Impedance graph.
Ctrl+F3	Open or select the Phase Response graph.
Ctrl+F4	Open or select the Group Delay graph.
Ctrl+G	Plot the crossover network / filter in all open graphs.

Shift+F1 to F7	Store the last plotted design into graph memory 1 to 7.
Shift+Ctrl+F1 to F8	Recall and replot memory 1 to 7. Use Shift+Ctrl+F8 to recall and replot all stored graph memories.
Ctrl+X	Clear the selected graph.
Ctrl+Y	Clear all open graphs.
Ctrl+U	Activate the cursor in the selected graph.
Ctrl+H	Hide the cursor in the selected graph.
Ctrl+I	Include or ignore the driver acoustic response.
Ctrl+Z	Include or ignore the net (composite) response.
Ctrl+C	Copy the selected graph to the Windows clipboard so it can be pasted into another Windows application as a picture (bitmap).
←	Move the graph cursor left one pixel.
→	Move the graph cursor right one pixel.
Shift+←	Move the graph cursor left 20 pixels.
Shift+→	Move the graph cursor right 20 pixels.
Ctrl+←	Switch the cursor left to the next filter plot line. For example, use Ctrl+← to move the cursor from the band-pass filter plot line to the low-pass filter plot line.
Ctrl+→	Switch the cursor right to the next filter plot line. For example, use Ctrl+→ to move the cursor from the band-pass filter plot line to the high-pass filter plot line.
↑	Move the graph cursor up through the last 10 design plots.
↓	Move the graph cursor down through the last 10 design plots.

Driver Database

Ctrl+W	Edit the driver information in the database.

A

Appendix B: Glossary of Terms

1-W SPL	The reference sensitivity of a driver with a 1 watt signal and measured at 1 meter as a sound pressure level (dBSPL).
2.8-V SPL	The reference sensitivity of a driver with a 2.83 volt signal and measured at 1 meter as a sound pressure level (dBSPL).
AC	Alternating current. Audio signals are AC signals.
aligned	The drivers in a multi-way speaker system are "aligned" when the sound emanating from each driver in the system arrives at a desired location in space at the same moment in time. Because the acoustical "center" of a driver changes with frequency, it is rarely possible to align drivers at all frequencies. Usually adjacent drivers are aligned at the crossover frequency.
amplitude response	A comparison of the magnitudes of an input and output signal versus frequency. It is commonly labeled the "frequency response" or "magnitude response". The output signal is usually measured as a pressure level.
APC	An all-pass crossover network with a flat amplitude response.
audible	Able to be heard. A sound that cannot be heard is inaudible. What makes a sound audible? Answer: 1) The frequency of the sound waves are within the range of human hearing. 2) The sound waves have sufficient amplitude (loudness) to be heard. 3) The sound waves are not "masked" by a louder sound or noise.
band-pass	A combined high-pass and low-pass filter that passes only a specified "band" of sound. Abbreviated "BP".
BL	The motor strength of a driver.
Bessel	A filter type with Q = 0.58 that produces a nearly flat amplitude response and group delay and a –5 dB crossover point.
Butterworth	A popular CPC filter type with Q = 0.707 that produces a –3 dB crossover point and a flat power response.
capacitance	A characteristic of an electric circuit that resists changes in voltage. One of its effects is to limit low-frequency signals in AC circuits. It is expressed in Farads (F).
capacitor	A device that stores electrical energy because its primary characteristic is capacitance. Capacitors also have resistance and inductance, but they are at much lower levels.
Chebychev	A seldom used filter type with Q = 1.0 that produces a 0 dB crossover point and approximate ±2 dB ripple in the amplitude response.
Cms	The mechanical compliance of the suspension of a driver or passive radiator.
coaxial	When two drivers are mounted like two wheels on an axle and face the same direction, they are said to be "coaxial"
coincidence	When two identical signals occupy the same point in space and time they are coincident and will sum to +6 dB if they have the same polarity. If one of them has

A

an inverted polarity, they will cancel each other. Two signals that are not coincident will usually sum to +3 dB.

coincident	When two drivers are constructed in a single frame such that their acoustical centers appear to be in the same location in space at their crossover frequency, they are said to be coincident.
coverage	The area to which a driver delivers uniform sound. Also called the "coverage area", it is measured in angles. For example, a horn may have a 90° horizontal coverage and a 40° vertical coverage. The shape of the coverage area is called the "coverage pattern". The coverage of most drivers decreases as the frequency increases.
CPC	A constant-power crossover network with a flat power response.
crossover point	Applies only to two or more drivers which have complimentary passbands that overlap. The crossover point is the frequency and level where the signals from the drivers intersect. For example, a crossover point may have a frequency of 500 Hz and a level of –6 dB.
crossover network	An electric circuit that "divides" the sound in a multi-way speaker. It is comprised of two or more filters such as low-pass, band-pass and high-pass filters.
DC	Direct current. DC flows in a steady direction. The direction depends on the polarity of the energy source. A battery produces a DC signal.
DCR	The DC resistance of an electric device such as an inductor.
DF	The damping factor. This can be the system damping factor in the case of a single-tuned bandpass box or the damping factor of an amplifier output.
Dia	The diaphragm or piston diameter of a driver or passive radiator.
diffraction	The bending of sound waves as they pass near an edge or corner of a solid object.
distortion	When a signal flowing through a circuit is compared at two different points, any change except for magnitude is distortion. Passive crossover network design is often concerned with "modulation" distortion. This is the distortion that results from the interaction of overlapping audio signals in the crossover region.
EBP	The efficiency bandwidth product (Fs/Qes). *Note: When Qes is unknown EBP is sometimes estimated with Fs/Qts.*
ESR	The equivalent series resistance of an electric device such as a capacitor.
F3	The half-power (–3 dB) frequency of a system. F3 is sometimes referred to as the "corner" or "cutoff" frequency.
Fb	The system resonant frequency of a speaker. It is also called the "tuning frequency" of a box with a vent. Double-tuned bandpass boxes have two Fb values.
FFT	Fast Fourier transform, a mathematical method of converting a continuous-time signal into a discrete-time signal so that it can be analyzed digitally.
Fill	The acoustic absorption or damping material added inside a box to suppress unwanted resonances (and sometimes to increase the apparent box volume).

frequency	The number of sound waves per second. It is expressed in hertz (Hz).
Fs	The free-air resonant frequency of a driver or passive radiator.
γ	(gamma) The ratio of heat at constant pressure to that at constant temperature for the air inside a box. It characterizes the change in the "springiness" of the air inside the box after acoustic absorption or "fill" is added inside. *Note: X•over Pro uses a delta (ratio of change) value for γ which equals 1 (one) for a box with no fill and increases as the amount of fill inside the box increases.*
Gaussian	A seldom used filter type that uses an asymmetrical filter topology and a –6 dB crossover point.
group delay	The delay of an output signal compared to the input signal. It is derived from the slope of the phase response and is expressed in milliseconds (msec).
high-pass	A filter that passes high frequencies and attenuates or "stops" low frequencies. Abbreviated "HP".
impedance	Any opposition to the flow of electricity. This opposition is from all sources, both resistive and reactive.
inductance	A characteristic of an electric circuit that resists changes in current. One of its effects is to limit high-frequency signals in AC circuits. It is expressed in Henries (H).
inductor	A device designed to create an electromotive force (emf) because its primary characteristic is inductance. Inductors also have resistance, but it is usually at a much lower level.
insertion loss	Inductors exhibit varying degrees of efficiency, dependent primarily on their DCR. The attenuation in dB of the signal through an inductor is its insertion loss.
isobaric	Constant pressure. A "compound" pair of drivers are mounted on either end of a small, sealed isobaric chamber. The pressure in the chamber is kept constant because both drivers are fed the same signal and their diaphragms move in the same direction.
Le	The inductance that a driver appears to have at upper frequencies because of the inductive reactance of its voice coil.
Legendre	A seldom used filter type that uses an asymmetrical filter topology and a –1 dB crossover point.
linear-phase	A seldom used filter type that uses an asymmetrical filter topology and a –6 dB crossover point.
Linkwitz-Riley	A popular APC filter type with a Q = 0.49 that produces a –6 dB crossover point and a maximally flat amplitude response. 4th-order Linkwitz-Riley filters are also called "squared Butterworth" filters.
low-pass	A filter that passes low frequencies and attenuates or "stops" high frequencies. Abbreviated "LP".
max flat	(maximally flat) An amplitude response curve with the least possible ripple in the passband.

midrange	A driver designed to produce frequencies in the middle of the audible spectrum (500 to 4000 Hz).
minimum phase	A system whose phase response is directly related to its amplitude response and that has no more phase shift other than what is required by the slope of its amplitude response curve.
misaligned	The drivers in a multi-way speaker are "misaligned" when the sound emanating from each driver in the system arrives at a different moment in time.
MLS	A pseudo-random maximum length sequence signal used for testing speakers and audio systems.
Mms	The mechanical mass of a driver or passive radiator diaphragm and voice coil assembly including the air load.
η_0	(eta zero) The reference efficiency of a driver with a half-space acoustical load.
nonpolarized	A device is nonpolarized if it passes a signal equally well in both directions. Nonpolarized components are required for audio circuits because audio signals are AC signals.
octave	A two-to-one change in frequency. For example, 100 to 200 Hz is one octave and 5000 to 10000 Hz is also one octave.
on-axis	When a speaker or driver is pointing directly at something it is "on-axis" to it. Most speaker measurements are made on-axis.
order	The "order" is a numerical filter classification that is determined by counting the total number of capacitor and inductor sections in the filter. (Exception: A band-pass filter is actually a low-pass and high-pass filter and so its total section count must be divided by two.) The order describes how fast a filter will attenuate sound in the stopband.
overhung	An overhung voice coil is taller than the height of the magnet gap.
parallel	Components that are connected side-by-side are wired in parallel.
passband	The operating frequency band of a driver. The passband is usually defined by F3.
Pe	The maximum electrical power that a driver can handle before it is damaged, usually when the voice coil burns. Also called the "thermal power limit".
phase	The change, measured as an angle of rotation, that an alternating current undergoes with the passage of time.
phase shift	When a signal flowing through a circuit is compared at two different points, any delay in the signal can be observed as a phase shift. It is expressed as an angle of rotation.
piston band	The frequency band where a driver maintains a constant load versus frequency.
pixel	The smallest dot of light that a computer can turn on and off on the video monitor.
polarity	A driver is said to be "in" polarity when its diaphragm moves outward in response to a positive signal. It is said to be invented or "out" of polarity when its diaphragm moves inward in response to a positive signal.

power response The total acoustic power emanating from a speaker or driver. It does not change with direction or distance from the source.

push-pull When two drivers are mounted in opposite directions and wired with opposite polarity with respect to each other, they are said to be in a "push-pull" configuration. This is because when the diaphragm of one driver moves away from its magnet, the other driver's diaphragm moves toward its magnet. This results in a reduction in even-order distortion because many nonlinearities are cancelled.

Q The resonance magnification of a system.

Qa The Q of a box resulting from all absorption losses. Absorption losses usually result from the addition of an acoustical absorber or "fill" to the box interior.

Qes The Q of a driver at Fs considering only its electrical (non-mechanical) resistance.

QL The Q of a box resulting from all leakage losses. Sources of leakage loss include box wall vibration, poor box construction, poor driver gasket seal, a porous driver dust cap and a "lossy" driver surround.

Qms The Q of a driver or passive radiator at Fs considering only its mechanical (non-electrical) resistance.

Qts The total Q of a driver at Fs considering both electrical and mechanical resistance.

Re The DC resistance of a driver's voice coil.

reactance An opposition to the flow of electricity because of capacitive and inductive characteristics. Purely resistive characteristics are not included in reactance.

resistance An opposition to the flow of electricity without capacitive and inductive characteristics.

resonance The frequency of peak response of a device that results from the balance of its capacitive and inductive characteristics.

Rms The mechanical resistance of a driver or passive radiator suspension losses.

RTA A real-time analyzer is an instrument that continuously measures the sound pressure level. It usually employs ⅓-octave band filters to analyze a "pink noise" signal and display an amplitude response with a bar graph.

Sd The diaphragm or piston area of a driver or passive radiator.

series Components that are connected end-to-end are wired in series.

slope The attenuation rate of a filter.

SPL The sound pressure level. It is usually expressed as a decibel (dB) ratio.

stopband The area outside of the operating frequency band (passband) of a filter.

subsonic Sound waves with such a low frequency that they cannot be heard and are therefore beyond the range of human hearing.

subwoofer A driver designed to produce ultra-low frequencies (below 100 Hz).

supertweeter A driver designed to produce ultra-high frequencies (above 5000 Hz).

T-S	The Thiele-Small driver or passive radiator parameters, named after A.N. Thiele and R.H. Small who popularized the "lumped sum" method of enclosure analysis used by many in the audio industry. Many of their key papers are included in the Audio Engineering Society *Loudspeaker Anthologies*, listed in Appendix E.
TDS	Time delay spectrometry is a patented measurement system invented by Richard C. Heyser which can make pseudo-anechoic measurements of speakers.
three-way	A crossover network with a low-pass, band-pass and high-pass filter to divide the sound three ways between a woofer, midrange driver and tweeter.
topology	The layout of a circuit, including the general pattern of connection and location of its various components.
tweeter	A driver designed to produce high frequencies (2000 to 20000 Hz).
two-way	A crossover network with a low-pass and high-pass filter to divide the sound two ways between a woofer and tweeter.
ultrasonic	Sound waves with such a high frequency that they cannot be heard and are therefore beyond the range of human hearing.
underhung	An underhung voice coil is shorter than the height of the magnet gap.
Vas	The volume of air having the same compliance or "springiness" as the suspension of a driver or passive radiator.
Vb	The net internal volume of a box.
Vd	The diaphragm or piston displacement volume of a driver or passive radiator. For a driver this is usually the volume displaced at Xmax. For a passive radiator this is the volume displaced at Xmech.
woofer	A driver designed to produce low frequencies (20 to 2000 Hz).
Xmax	The maximum linear excursion of the driver or passive radiator. It should be measured in one direction from a resting position.
Xmech	The maximum mechanical excursion of the driver or passive radiator. With some drivers Xmech is reached when the driver's diaphragm has moved as far as the suspension will allow. In other drivers Xmech is reached when the voice coil former hits the back plate of the magnet structure. Xmech should be measured in one direction from a resting position.
Z	The nominal electromagnetic impedance of a driver.
Zt	The total or net load impedance. This is the load impedance that a corresponding low-pass, band-pass or high-pass crossover network filter will "see". If an impedance equalization network is present then the load impedance should remain relatively constant with frequency and crossover filter design will be relatively predictable. If an impedance equalization network is not present, the impedance may vary with frequency and Zt should be considered a nominal value only (much like the driver's nominal impedance, Z).

Appendix C: Open Back Driver Shapes

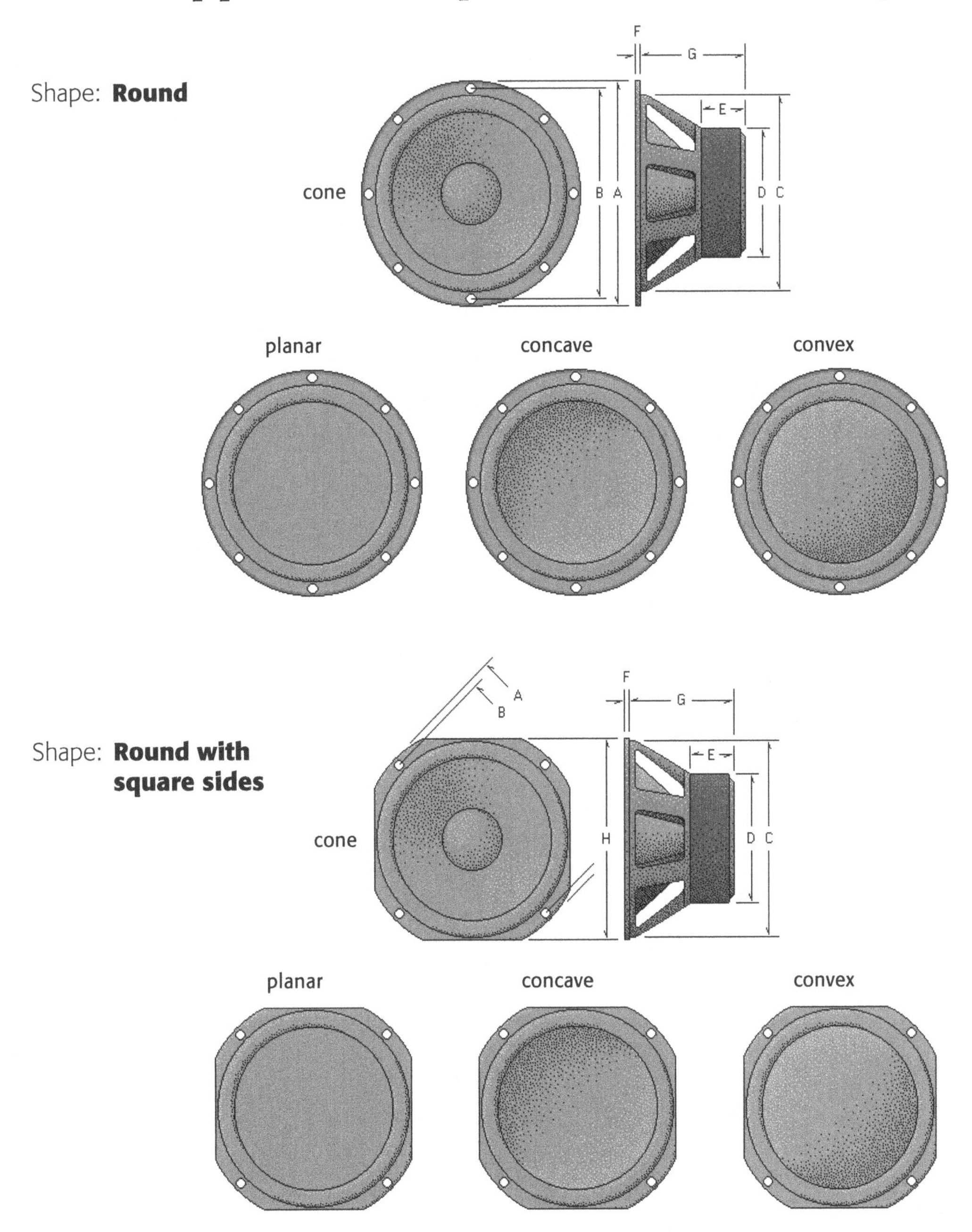

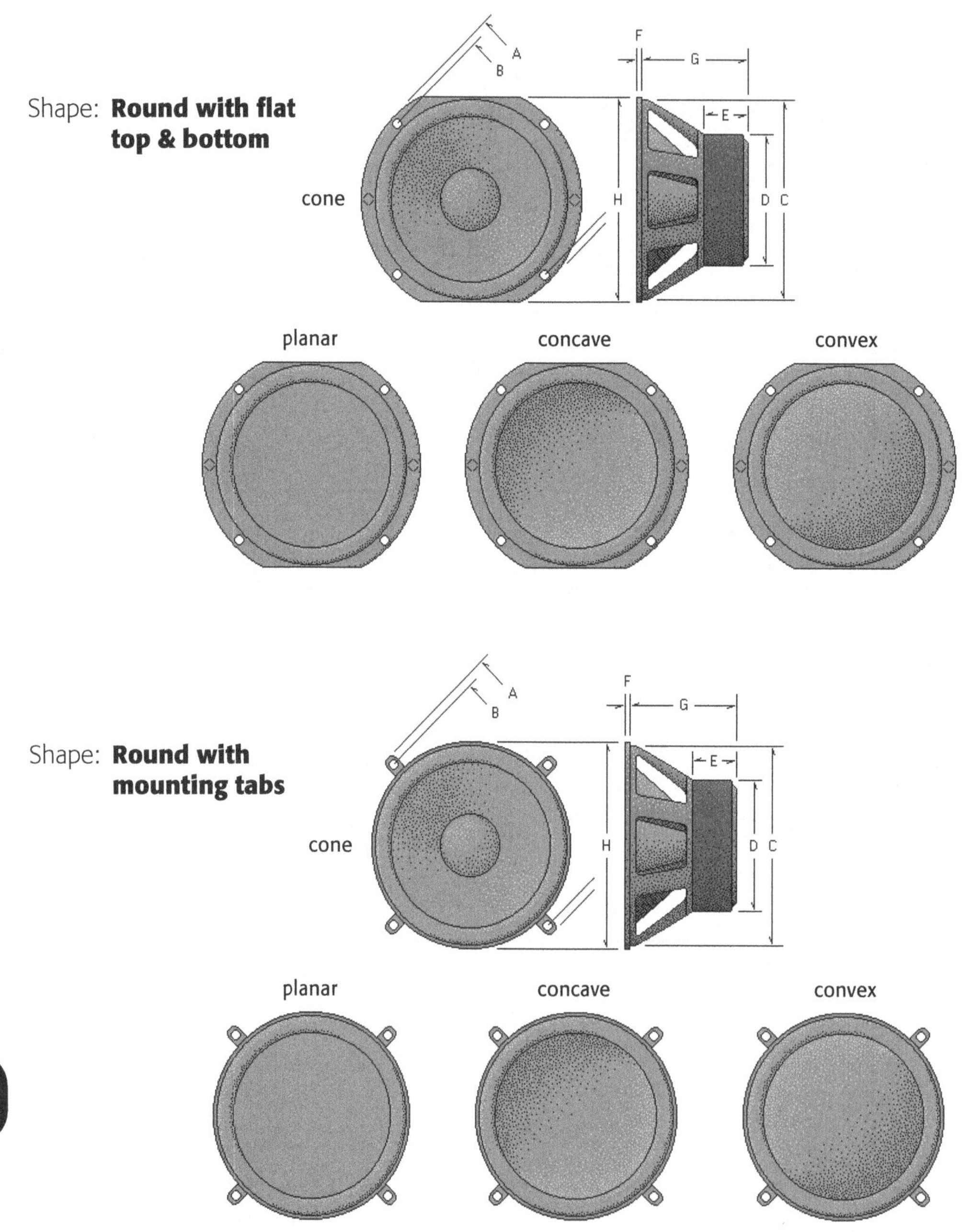
Shape: **Round with flat top & bottom**
cone
A
B
F
G
E
H
D C
planar
concave
convex
Shape: **Round with mounting tabs**
cone
A
B
F
G
E
H
D C
planar
concave
convex

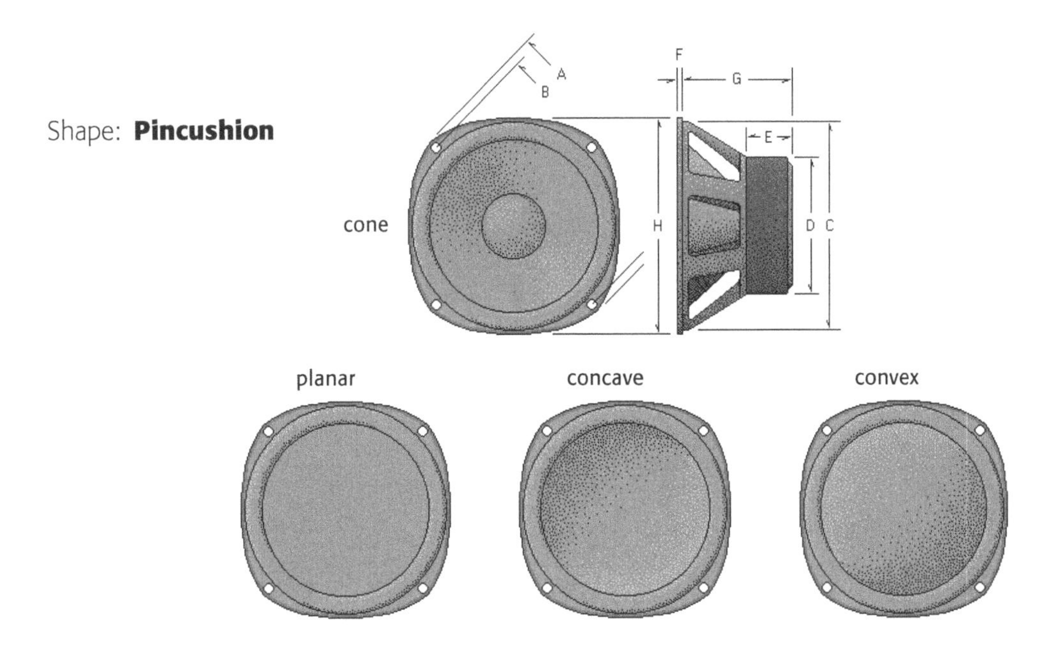
Shape: **Pincushion**
cone
A
B
F
G
E
H
D
C
planar
concave
convex

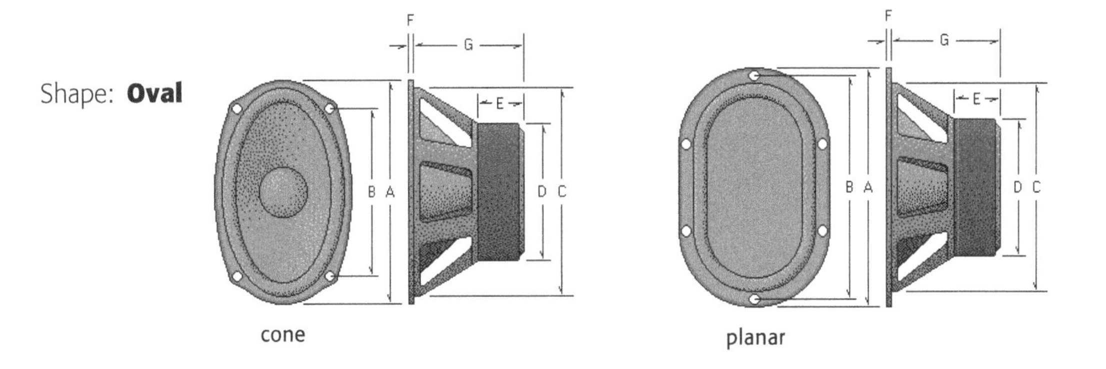
Shape: **Oval**
F
G
E
B
A
D
C
cone
planar

Shape: **Square**

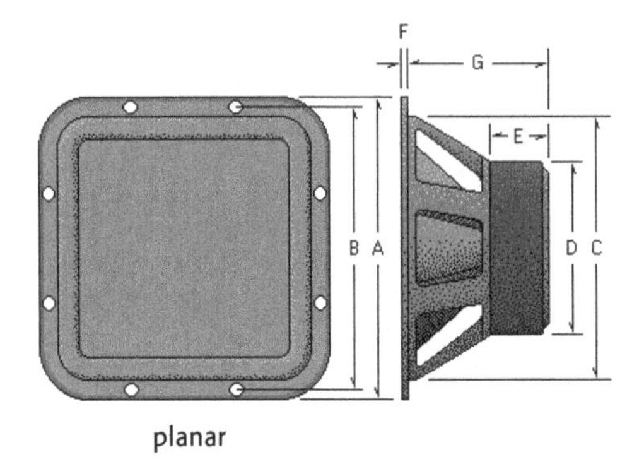

planar

Shape: **Rectangular**

F
G
E
B A
D C

cone

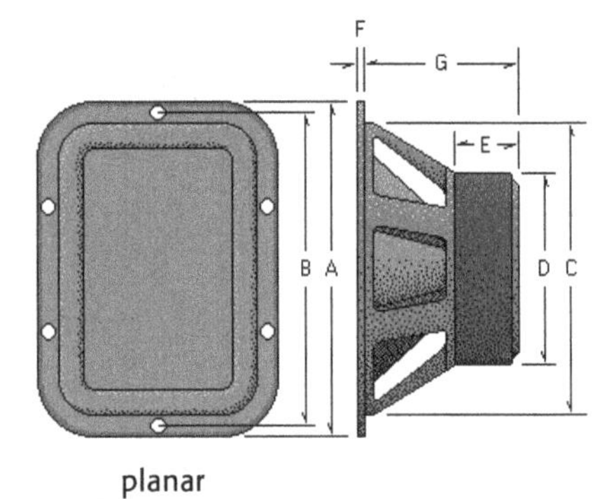

planar

A

Appendix D: Sealed Back Driver Shapes

Shape: **Round**

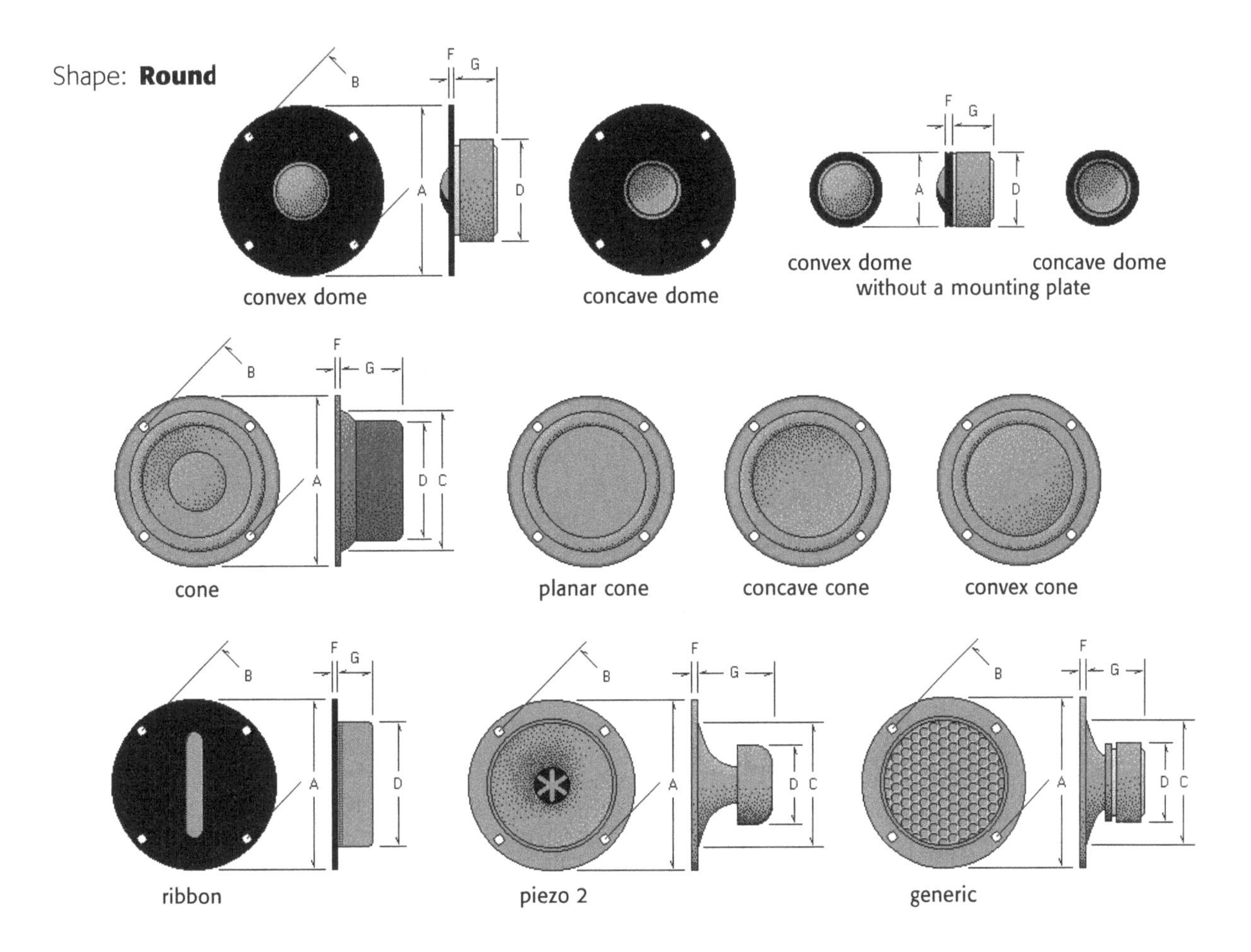

Shape: **Round with square sides**

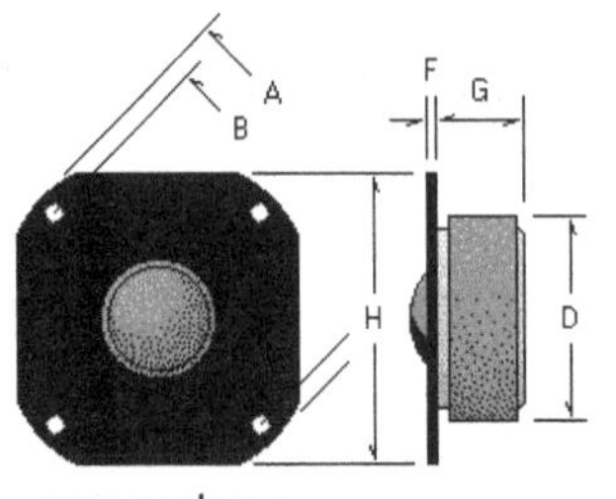
convex dome

concave dome

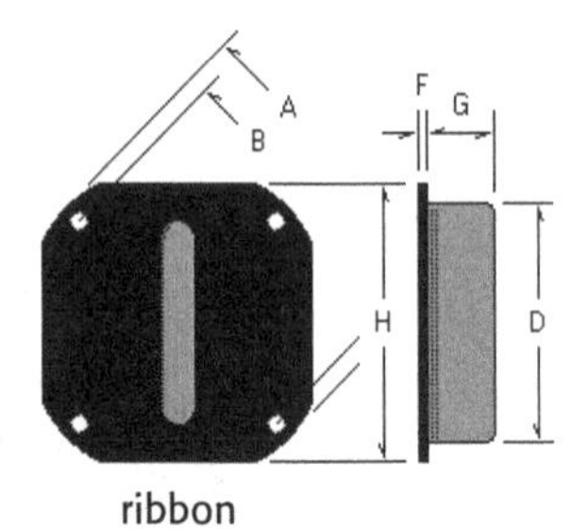
ribbon

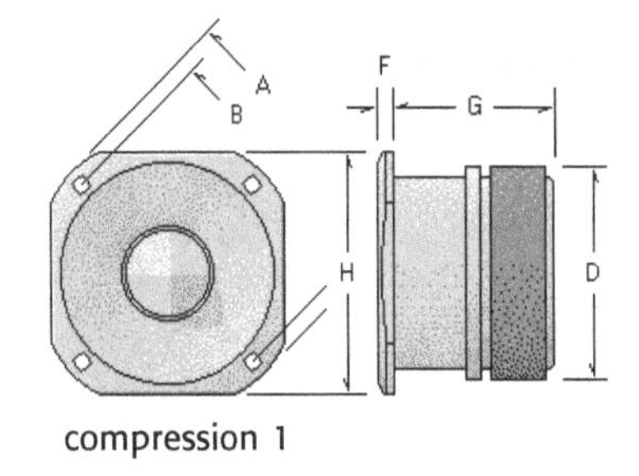
compression 1

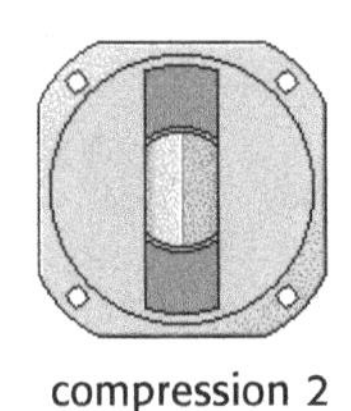
compression 2

Shape: **Round with one flat side**

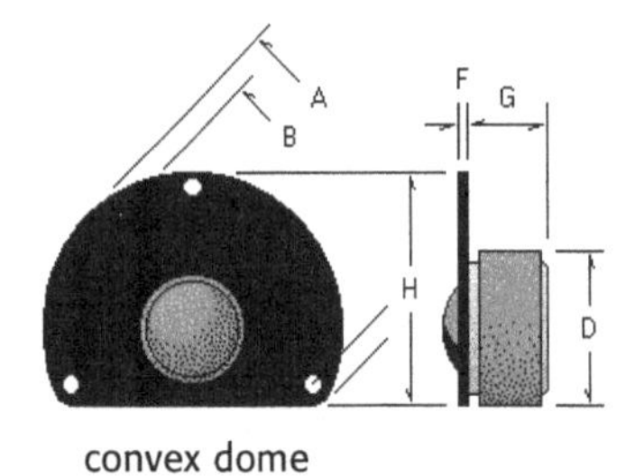
convex dome

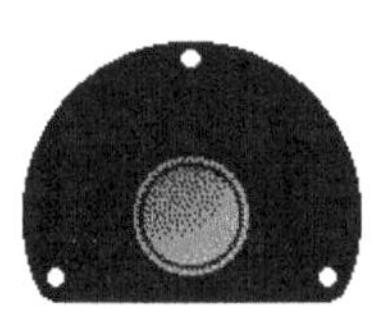
concave dome

Shape: **Round with two flat sides**

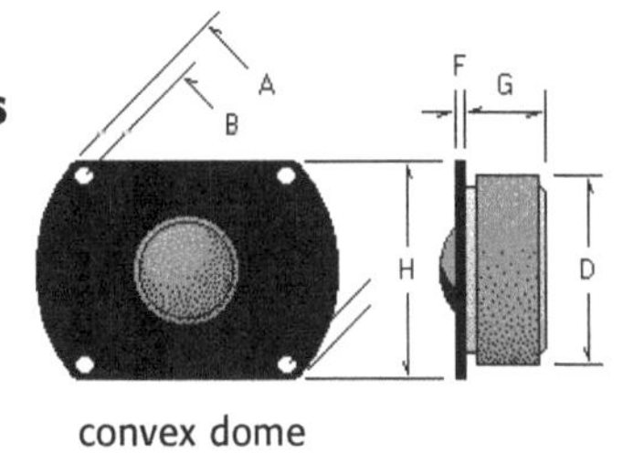
convex dome

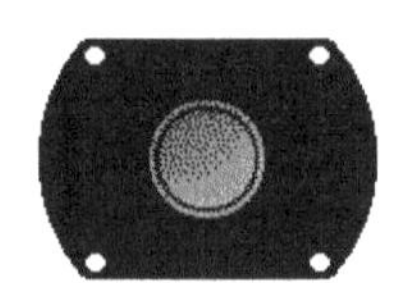
concave dome

Shape: **Round with mounting tabs**

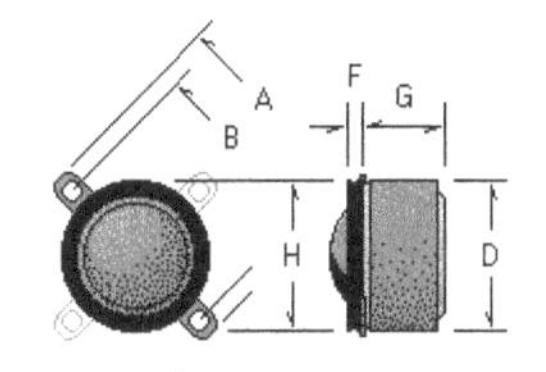

convex dome

concave dome

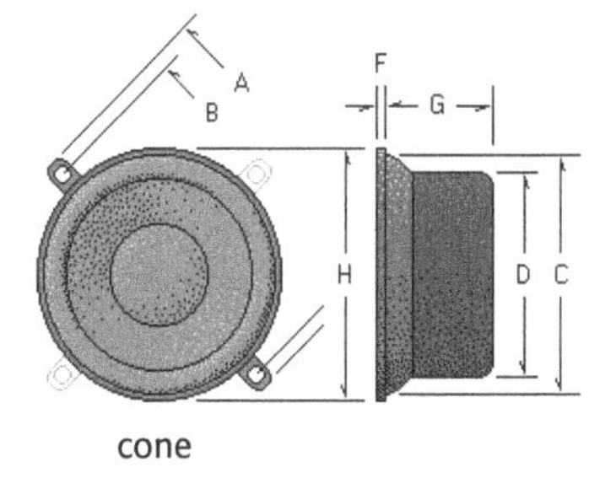

cone

planar cone

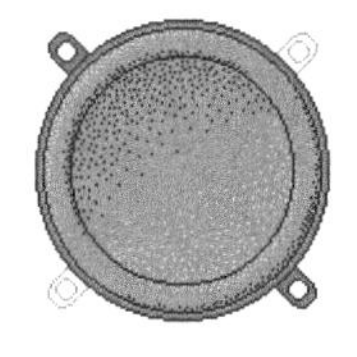

concave cone

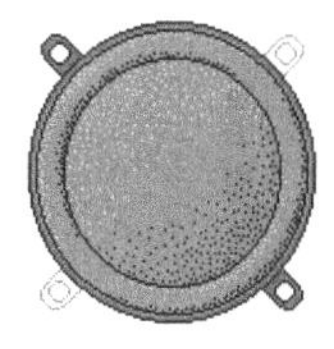

convex cone

Shape: **Pincushion**

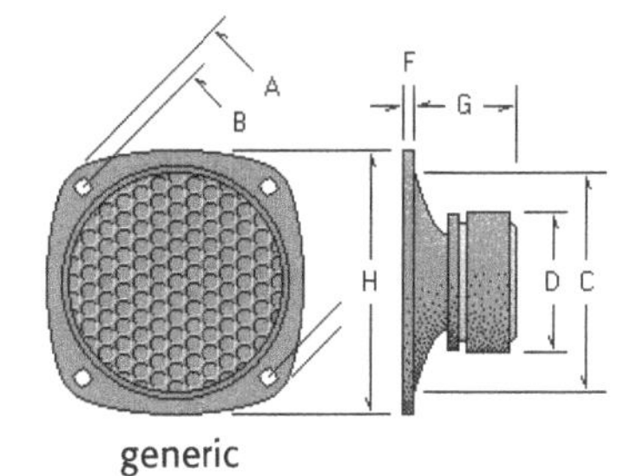

generic

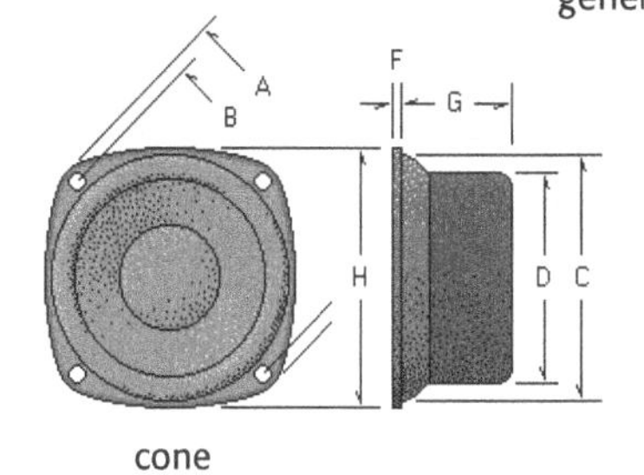

cone

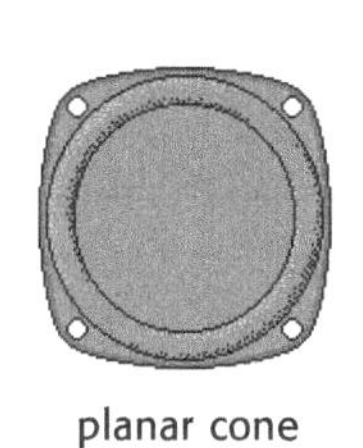

planar cone

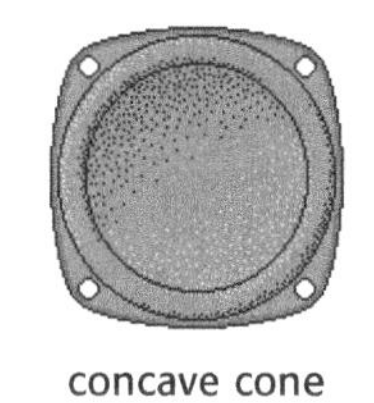

concave cone

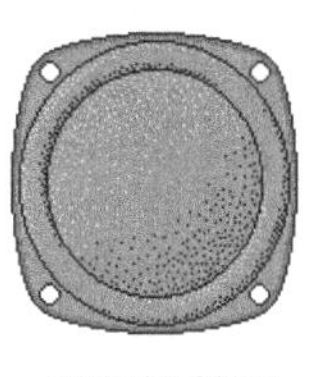

convex cone

A

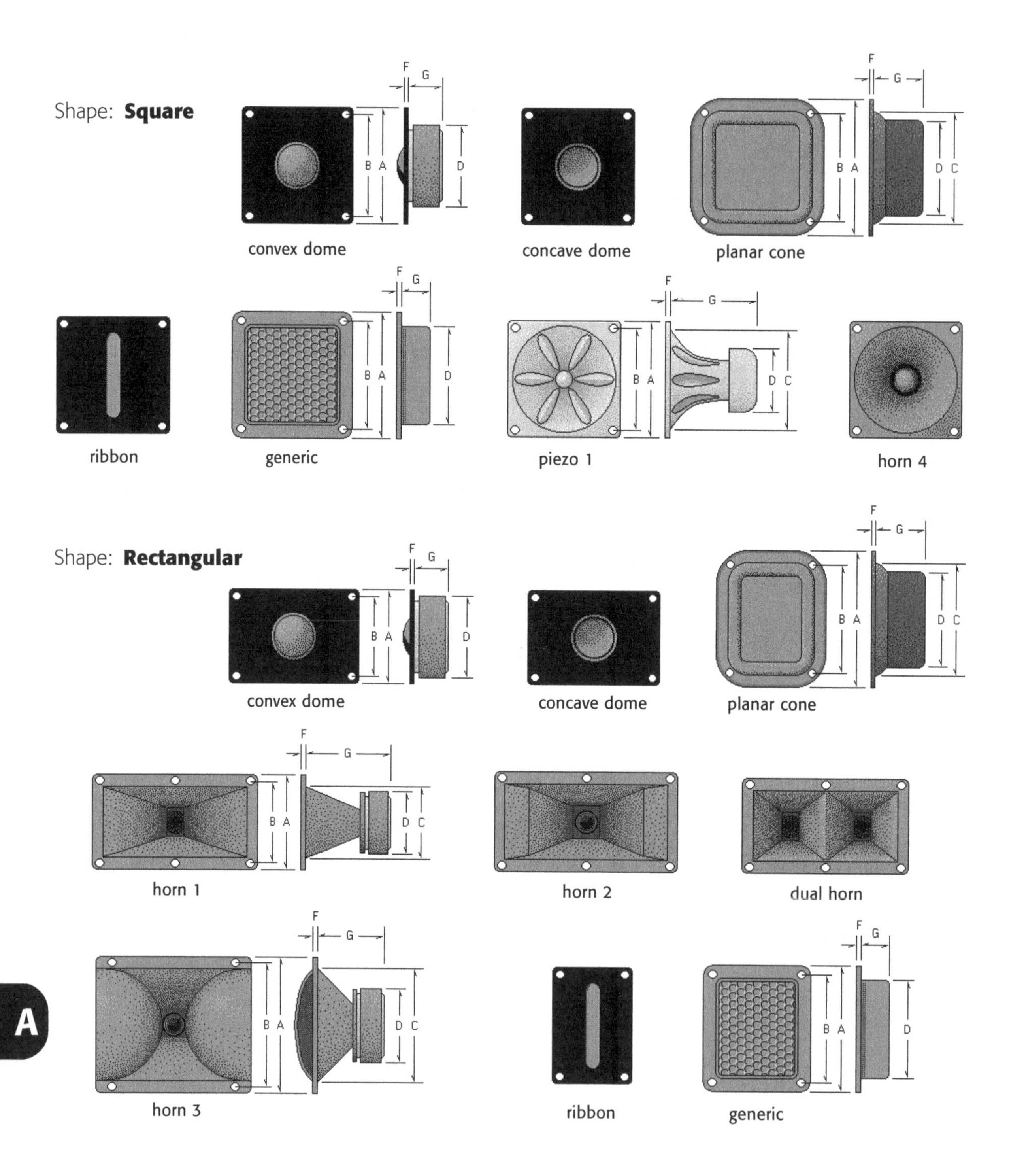
Shape: Square
F
G
B
A
D
C
convex dome
concave dome
planar cone
ribbon
generic
piezo 1
horn 4
Shape: Rectangular
convex dome
concave dome
planar cone
horn 1
horn 2
dual horn
horn 3
ribbon
generic
A

Appendix E: Suggested Reading

Beginner-Level Reading
Book: *Designing, Building & Testing Your Own Speaker System* by David B. Weems.
Pennsylvania: Tab Books Inc. (McGraw-Hill, Inc.), 1981, 1990 (revised edition)

Book: *How to Build Speaker Enclosures* by A. Badmaieff and D. David
Indiana: Howard W. Sams & Co., Inc., 1966

Intermediate-Level Reading
Book: *The Loudspeaker Design Cookbook*, 5th Edition by Vance Dickason
New Hampshire: Audio Amateur Press, 1995

Book: *Advanced Speaker Designs for the Hobbyist and Technician* by Ray Alden
Indiana: Prompt Publications (Howard W. Sams & Co., Inc.), 1995

Book: *High Performance Loudspeakers*, 5th Edition by Martin Colloms
England: John Wiley & Sons, 1997

Chapter: "Loudspeakers, Enclosures, and Headphones" by Clifford A. Henricksen
Book: *Handbook for Sound Engineers*
Indiana: Howard W. Sams & Co., Inc., 1987

Advanced-Level Reading
Book: *Loudspeakers, An Anthology, Volume 1*, 2nd Edition
(Vol. 1 – Vol. 25 of *Journal of the Audio Engineering Society*)
New York: Audio Engineering Society, 1980

Book: *Loudspeakers, An Anthology, Volume 2*
(Vol. 26 – Vol. 31 of *Journal of the Audio Engineering Society*)
New York: Audio Engineering Society, 1984

Book: *Theory and Design of Loudspeaker Enclosures* by J.E. Benson
Indiana: Synergetic Audio Concepts, 1993

Book: *Testing Loudspeakers*, 1st Edition by Joseph D'Appolito
New Hampshire: Audio Amateur Press, 1998

Additional Reading & References
Article: "Thiele, Small, and Vented Loudspeaker Design: Parts I–IV" by Robert M. Bullock III
Magazine: Issues Four/80 through Three/81 of *Speaker Builder*
New Hampshire: Audio Amateur Publications, 1980-81

A

Article: "Double Chamber Speaker Enclosure" by George L. Augspurger
Magazine: Dec. Issue/1961 of *Electronics World Magazine*
Ziff-Davis Publishing Co., 1961

Article: "New Guidelines for Vented-Box Construction" by George L. Augspurger
Magazine: Issue Two/91 of *Speaker Builder*
New Hampshire: Audio Amateur Publications, 1991

Article: "The Third Dimension: Symmetrically Loaded, Part 1" by Jean Margerand
Magazine: Issue Six/88 of *Speaker Builder*
New Hampshire: Audio Amateur Publications, 1988

Article: "Passive Crossover Networks: Parts 1–3" by Robert M. Bullock III
Magazine: Issues One/85–Three/85 of *Speaker Builder*
New Hampshire: Audio Amateur Publications, 1985

Paper: "A Bandpass Loudspeaker Enclosure" by L.R. Fincham
Preprint 1512 presented at the 63rd Convention of the Audio Engineering Society
Journal of the Audio Engineering Society (Abstracts), Vol. 27, p. 600 (1979 July/Aug.)

Paper: "An Introduction to Band-Pass Loudspeaker Systems" by Earl R. Geddes
Publication: Vol. 37, No. 5 of *Journal of the Audio Engineering Society*
New York: Audio Engineering Society, 1989

Paper: "Impedance Analysis of Subwoofer Systems" by Arthur P. Berkhoff
Publication: Vol. 42, No. 1/2 of *Journal of the Audio Engineering Society*
New York: Audio Engineering Society, 1994

Paper: "Passive-Radiator Loudspeaker Systems Part 1: Analysis" by Richard H. Small
Publication: Vol. 22, No. 8 of *Journal of the Audio Engineering Society*
New York: Audio Engineering Society, 1974

Paper: "Passive-Radiator Loudspeaker Systems Part 2: Synthesis" by Richard H. Small
Publication: Vol. 22, No. 9 of *Journal of the Audio Engineering Society*
New York: Audio Engineering Society, 1974

Article: "Matching Driver Efficiencies" by John I. Lipp
Magazine: Issue Five/93 of *Speaker Builder*
New Hampshire: Audio Amateur Publications, 1993

Paper: "Complete Response Function and System Parameters for a Loudspeaker with Passive Radiator" by Douglas H. Hurlburt
Publication: Vol. 48, No. 3 of *Journal of the Audio Engineering Society*
New York: Audio Engineering Society, 2000

Appendix F: Driver Parameter Worksheet

The following worksheet lists the driver parameters used by this program. Photocopy it and use it whenever you need to collect the parameters for unknown drivers.

General Information

❑ Open Back ❑ Sealed Back

Manufacturer: ____________________

Model Name: ____________________

Serial No: ____________________

Comment: ____________________

____

____

____

❑ Underhung Voice Coil ❑ Magnetic Shielding ❑ Ferrofluid

❑ Dual Voice Coil

Mechanical Parameters	*Units*
Fs: ________	Hz
Qms: ________	
Vas: ________	________
Cms: ________	________
Mms: ________	________
Rms: ________	________
Xmax: ________	________
Xmech: ________	________
Dia: ________	________
Sd: ________	________
Vd: ________	________

Electrical Parameters	*Units*	*Parallel Dual VC*	*Series Dual VC*
Qes: ________		Qes: ________	Qes: ________
Re: ________	ohms	Re: ________	Re: ________
Le: ________	millihenries	Le: ________	Le: ________
Z: ________	ohms	Z: ________	Z: ________
BL: ________	________	BL: ________	BL: ________
Pe: ________	watts	Pe: ________	Pe: ________

Combination Parameters	*Units*		
Qts: ________		Qts: ________	Qts: ________
ηo: ________	%	ηo: ________	ηo: ________
1-W SPL: ________	dB	1-W SPL: ________	1-W SPL: ________
2.8-V SPL: ________	dB	2.8-V SPL: ________	2.8-V SPL: ________

A

Appendix G: Acoustic Response Worksheet

The following worksheet lists the data points used by this program for the measured driver acoustic response. Photocopy it and use it whenever you need to manually collect the response data.

Hz	*Level (dB)*	*Hz*	*Level (dB)*	*Hz*	*Level (dB)*	*Hz*	*Level (dB)*	*Hz*	*Level (dB)*	*Hz*	*Level (dB)*
5		36		135		500		1.9 k		7 k	
6		38		140		530		2 k		7.5 k	
8		40		150		560		2.1 k		8 k	
10		42		160		600		2.25 k		8.5 k	
11		45		170		630		2.35 k		9 k	
12		47		180		670		2.5 k		9.5 k	
13		50		190		700		2.65 k		10 k	
14		53		200		750		2.8 k		10.5 k	
15		56		210		800		3 k		11 k	
16		60		225		850		3.15 k		12 k	
17		63		235		900		3.35 k		12.5 k	
18		67		250		950		3.55 k		13.5 k	
19		70		265		1 k		3.75 k		14 k	
20		75		280		1.05 k		4 k		15 k	
21		80		300		1.1 k		4.2 k		16 k	
22		85		315		1.2 k		4.45 k		17 k	
24		90		335		1.25 k		4.75 k		18 k	
25		95		355		1.35 k		5 k		19 k	
27		100		375		1.4 k		5.3 k		20 k	
28		105		400		1.5 k		5.6 k			
30		110		420		1.6 k		6 k			
32		120		445		1.7 k		6.3 k			
34		125		475		1.8 k		6.7 k			

Index

Made in the USA
Columbia, SC
05 February 2018